Contents

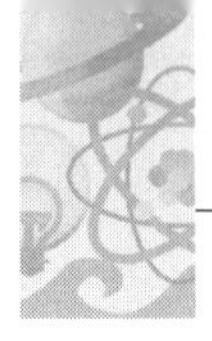

Introduction

About this guide

This unit guide covers essential material for the OCR A2 Physics A specification. It contains material needed for the effective study of **Unit G484: The Newtonian World**. This unit is assessed by a 1 hour written examination. The aim of the guide is to help you *understand* the physics, so that you can effectively *revise* and *prepare* for the examination.

The **Content Guidance** section is based on the structure of the specification. The same headings are used and the sub-headings closely follow the learning outcomes of the specification. This guide is not intended to be a detailed textbook and does not contain every fact that you need to know. The focus is on understanding the principles and definitions so that you can successfully tackle the varied questions in the examination. This guide uses worked examples to illustrate good practice and offers an examiner's perspective on how you can improve your answers.

The **Question and Answer** section shows the type of questions you can expect in the unit examination. The questions are illustrated with answers given by typical C-grade and A-grade candidates. The answers are followed by comments from the examiner explaining why marks are awarded or lost. Common errors made by candidates are also highlighted, so that you do not make the same mistakes. Candidates frequently lose marks for careless mistakes, incomplete answers, muddled presentation and illegible handwriting.

Use of this guide

This guide can be used throughout the course and not just at revision time. The content of the guide is set out in the order of the learning outcomes of the specification, so that you can use it:

- to check your notes
- as a reference for internal tests and homework
- to revise in manageable sections
- to identify your strengths and weaknesses
- to check that you have covered the specification completely
- to learn the definitions expected by the examiners
- to improve the quality of your answers
- to familiarise yourself with the range of questions you can expect in the examination
- to improve your confidence in applying physics

Revision

Planning study and revision is essential. You can avoid disappointment and anxiety by organising a revision plan well before your examination. You cannot cram in one year's physics into a couple of days. Some points to consider when you are revising for an examination are given below.

- Always start with a topic that you find easy. This will boost your confidence.
- Learn all the equations listed in the specification.
- Learn equations, definitions and laws in the specification thoroughly. Examiners expect perfection with definitions and laws and tend to give no marks for using wrong equations.
- Make your revision active by writing out the equations, laws and definitions, drawing diagrams and doing many calculations.
- Write a brief summary of the topic.
- Ask your teacher to explain any words, laws or definitions that do not make sense.
- When revising, make sure you have your notes, the specification and a reference book available to complement this guide.
- Make good use of the specification. Use a highlighter pen to identify the topics that you have already covered.
- Do not revise for too long. If you are tired, you cannot produce quality work.
- Do not leave your revision to the last moment. Plan out a strategy spread over many weeks before the actual examination. Work hard during the day and learn to relax when needed.

Key skills

In examinations, candidates lose many marks because of their inability to apply some basic skills.

Calculator work

Some common mistakes are highlighted below.

- $\frac{3+4}{0.5}$ is put into the calculator as [3] [+] [4] [÷] [0.5], which gives 11. It should be [(] [3] [+] [4] [)] [÷] [0.5], which gives the correct answer of 14.
- $\frac{4}{10 \times 2}$ is put into the calculator as [4] [÷] [10] [×] [2], which gives 0.8. It should be [4] [÷] [(] [10] [×] [2] [)] or [4] [÷] [10] [÷] [2], which gives the correct answer of 0.2.
- 3.0×10^8 is put into the calculator as [3.0] [×] [10] [EXP] [8], which gives 3.0×10^9. It should be [3.0] [EXP] [8]. (On some calculators, the 'exponent' button is 10^x.)
- Forgetting to insert the minus sign for the powers of 10.
- Squares are not put in when they are needed.
- Using lg (log to the base 10) instead of ln (log to the base e) or vice versa.

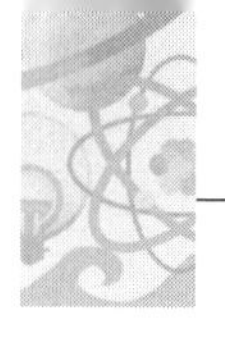

Numerical work

Some common mistakes are highlighted below.

- An inappropriate number of *significant figures* is used. If the data in a question are given to two significant figures, your answer must also be given to two significant figures. Examiners tend not to penalise an answer with too many significant figures. In fact, candidates lose more marks by using too few significant figures.
- Rounding up answers can lead to an incorrect answer. It is best to keep all the significant figures on your calculator as you progress through a calculation.
- Significant figures and decimal places may be confused. For example, 0.0254 is a number given to three significant figures, written to four decimal places. Using standard form, it is clear that 2.54×10^{-2} is written to three significant figures.
- Answers are not checked to see whether they are sensible. For example, the mass of a car is determined to be −600 kg. A negative mass is impossible. This answer should prompt you to check for an error in the calculation.

Algebraic work

Candidates lose marks by not being able to complete some simple algebraic operations. Here are some rules worth knowing:

- $a = b + c$ hence, $b = a - c$
- $a = bc$ hence, $b = \frac{a}{c}$
- $a = \frac{b}{c}$ hence, $b = ac$ and $c = \frac{b}{a}$
- $a^2 = b$ hence, $a = \pm\sqrt{b}$
- $y = e^x$ hence $x = \ln(y)$

Descriptive work

In examinations, candidates tend to gain more marks from mathematical questions than questions requiring descriptive answers. Descriptive questions are often answered badly for the following reasons:

- The answer is of an inappropriate length. Keep an eye on the number of marks available for a written answer and write accordingly. Answers that are too long waste time and may be repetitive; answers that are too short almost inevitably miss out key points that the examiner is looking for.
- Lack of physics in the answer. You are expected to state and use physics principles and vocabulary when answering questions requiring extended writing.
- The structure of the written answer is poor. Try not to ramble. Think carefully about the physics you want to put down on paper.
- Poor construction of sentences and bad spelling. Some questions in the examination paper have marks reserved for quality of written communication. You need to be careful when answering these questions.
- The question is misinterpreted. Read the question carefully and only start writing the answer once you are sure that you have understood what the examiner wants you to do.

Answering examination questions

Examiners do not set questions to trick you. They simply want you to demonstrate the extent of your knowledge and understanding of physics. Examiners spend much time discussing the wording of questions and their aim is to give you all the information succinctly so that you do not have to waste time deciphering the question.

When answering a question, you are expected to present your ideas logically. In an extended writing question, present your ideas in clear steps that show good use of your physics knowledge. In a calculation, clearly show all your working, including:

- the correct equation
- correct substitution in the right units
- correct algebraic manipulation
- correct answer with correct units

Do not waste your time in the examination. To get full marks you need to achieve, on average, 1 mark every minute. Read the whole of each question before you put pen to paper. Make sure that all calculations are done in the correct units and you have taken account of all prefixes such as 'milli', 'kilo' etc. Try to check your work as you go along. It is not sensible to check all your answers just at the end of the paper because you will have forgotten the finer details of each question. You should, however, do a quick check for units and significant figures once you have finished the paper, if you have some spare time at the end.

Many candidates waste time drawing diagrams using rulers. There are no extra marks for using a straight edge. In an examination, most diagrams can be drawn freehand. There is no point drawing a circuit diagram with the skills of a draughtsman when a freehand sketch showing all the components will do. Learn to save time in an examination.

In the Question and Answer section of this guide, there are many comments on the mistakes candidates make. As you get closer to your examination, make sure that you read the examiner's comments at the end of each question.

Assessing 'stretch and challenge' and synopticity

Some of the A2 exam questions, referred to as 'stretch and challenge' questions, are demanding. These questions are devised for candidates hoping to attain the new A* grade. Such questions may require

- connections to be made between different topics in physics
- open-ended responses
- extended writing
- greater mathematical competence

All A2 exam papers also assess your synoptic understanding of physics. In the Unit G484 exam paper, a small number of marks is reserved for the application of ideas from the AS units of Mechanics (G481) and Electrons, Waves and Photons (G482).

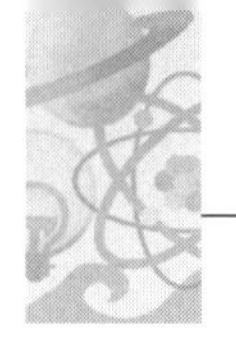

Command words used in examinations

The list below shows the most frequently used command words and their meanings:

- **Calculate** — this is used when a numerical answer is required. Show all your working and give an appropriate unit for your final answer. The number of significant figures must reflect the given data.
- **Define** — a formal statement or a word equation is required. Do not use symbols unless you are prepared to define them.
- **Deduce** — you have to draw conclusions using the information provided.
- **Describe** — this requires you to state in words, using diagrams if appropriate, the main points of the topic. The amount of description depends on the mark allocation. Bullet points are acceptable unless marks are being awarded for quality of written communication.
- **Determine** — use the information available to calculate the quantity required.
- **Estimate** — this requires a calculation in which you make sensible assumptions and use realistic values for quantities. Always check whether the final answer gives a realistic value.
- **Explain** — you have to use the correct physics vocabulary and principles. The depth of your answer depends on the number of marks available.
- **Select** — you will be given a list of key equations. Make sure you choose the most appropriate equations to do your calculations.
- **Show** — the answer to a particular problem is given in the question and is required in a subsequent question. You need to show each stage of your working in order to pick up all the available marks. This is not a calculator exercise, but you can use a calculator to help you reach the correct answer.
- **Sketch** — a simple freehand drawing is required. Significant detail should be added and labelled clearly.
- **Sketch a graph** — the shape of the graph needs to be correct. You may be expected to show values of the intercepts or the gradient. Axes must be fully labelled and the origin shown, if appropriate.
- **State** — you are expected to write a brief answer without any supporting or justifying statements.
- **Suggest** — there is often no single correct answer. You will be given credit for sensible reasoning based on correct physics.

Data, formulae and relationships

You will be given the following information when you take the examination for Unit G484: The Newtonian World.

Data

Values are given to three significant figures, except where more are useful.

Speed of light in a vacuum	c	$3.00 \times 10^{8}\,m\,s^{-1}$
Permittivity of free space	ε_0	$8.85 \times 10^{-12}\,C^2\,N^{-1}\,m^{-2}\,(F\,m^{-1})$
Elementary charge	e	$1.60 \times 10^{-19}\,C$
Planck constant	h	$6.63 \times 10^{-34}\,J\,s$
Gravitational constant	G	$6.67 \times 10^{-11}\,N\,m^2\,kg^{-2}$
Avogadro constant	N_A	$6.02 \times 10^{23}\,mol^{-1}$
Molar gas constant	R	$8.31\,J\,mol^{-1}\,K^{-1}$
Boltzmann constant	k	$1.38 \times 10^{-23}\,J\,K^{-1}$
Electron rest mass	m_e	$9.11 \times 10^{-31}\,kg$
Proton rest mass	m_p	$1.673 \times 10^{-27}\,kg$
Neutron rest mass	m_n	$1.675 \times 10^{-27}\,kg$
Alpha particle rest mass	m_α	$6.646 \times 10^{-27}\,kg$
Acceleration of free fall	g	$9.81\,m\,s^{-2}$

Conversion factors

Unified atomic mass unit	$1\,u = 1.661 \times 10^{-27}\,kg$
Electronvolt	$1\,eV = 1.60 \times 10^{-19}\,J$
Time	$1\,day = 8.64 \times 10^{4}\,s$
	$1\,year \approx 3.16 \times 10^{7}\,s$
Distance	$1\,light\,year \approx 9.5 \times 10^{15}\,m$

Mathematical equations

Arc length = $r\theta$

Circumference of circle = $2\pi r$

Area of circle = πr^2

Curved surface area of cylinder = $2\pi rh$

Volume of cylinder = $\pi r^2 h$

Curved surface area of sphere = $4\pi r^2$

Volume of sphere = $\frac{4}{3}\pi r^3$

Pythagoras' theorem: $a^2 = b^2 + c^2$

For small angles: $\theta \Rightarrow \sin\theta \approx \tan\theta \approx \theta$ and $\cos\theta \approx 1$

$\lg(AB) = \lg(A) + \lg(B)$

$\lg\left(\frac{A}{B}\right) = \lg(A) - \lg(B)$

$\ln(x^n) = n\ln(x)$

$\ln(e^{kx}) = kx$

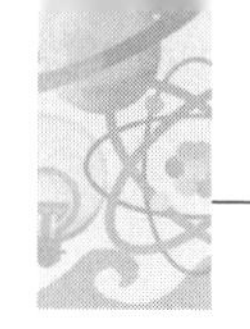

Formulae and relationships

Listed below are the formulae and relationships that you are given for Unit G484. They include the formulae and relationships that you were given for Units G481 and G482.

Unit G481: Mechanics

$F_x = F\cos\theta$

$F_y = F\sin\theta$

$a = \frac{\Delta v}{\Delta t}$

$v = u + at$

$s = \frac{1}{2}(u + v)t$

$s = ut + \frac{1}{2}at^2$

$v^2 = u^2 + 2as$

$F = ma$

$W = mg$

moment $= Fx$

torque $= Fd$

$\rho = \frac{m}{V}$

$p = \frac{F}{A}$

$W = Fx\cos\theta$

$E_k = \frac{1}{2}mv^2$

$E_p = mgh$

efficiency $= \frac{\text{useful energy output}}{\text{total energy input}} \times 100\%$

$F = kx$

$E = \frac{1}{2}Fx$

$E = \frac{1}{2}kx^2$

stress $= \frac{F}{A}$

strain $= \frac{x}{L}$

Young modulus $= \frac{\text{stress}}{\text{strain}}$

Unit G482: Electrons, Waves and Photons

$\Delta Q = I\Delta t$

$I = Anev$

$W = VQ$

$V = IR$

$R = \frac{\rho L}{A}$

$R = R_1 + R_2 + \dots$

$\frac{1}{R} = \frac{1}{R_1} + \frac{1}{R_2} + \dots$

$P = VI \quad P = I^2R \quad P = \frac{V^2}{R}$

$W = VIt$

e.m.f. $= V + Ir$

$V_{out} = \frac{R_2}{R_1 + R_2} \times V_{in}$

$v = f\lambda$

$\lambda = \frac{ax}{D}$

$d\sin\theta = n\lambda$

$E = hf \quad E = \frac{hc}{\lambda}$

$hf = \phi + KE_{max}$

$\lambda = \frac{h}{mv}$

Unit G484: The Newtonian World

$F = \frac{\Delta p}{\Delta t}$

$v = \frac{2\pi r}{T}$

$a = \frac{v^2}{r}$

$F = \frac{mv^2}{r}$

$F = -\frac{GMm}{r^2}$

$g = \frac{F}{m}$

$g = -\frac{GM}{r^2}$

$T^2 = \left(\frac{4\pi^2}{GM}\right)r^3$

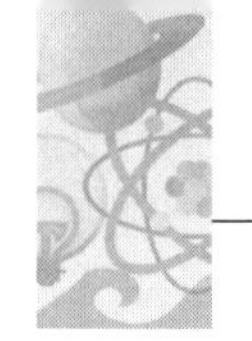

$$f = \frac{1}{T}$$

$$\omega = \frac{2\pi}{T} = 2\pi f$$

$$a = -(2\pi f)^2 x$$

$$x = A\cos(2\pi ft)$$

$$v_{max} = (2\pi f)A$$

$$E = mc\Delta\theta$$

$$pV = NkT$$

$$pV = nRT$$

$$E = \frac{3}{2}kT$$

This section is a student's guide to the A2 Unit G484: The Newtonian World. It covers all the relevant key facts, explains the essential concepts and highlights common misconceptions. The main topics are:

- Newtonian laws of motion
- Collisions
- Circular motion
- Gravitational fields
- Simple harmonic motion
- Solids, liquids and gases
- Temperature
- Thermal properties of materials
- Ideal gases

Newtonian laws of motion

Newton's first law

Newton's laws provide the basis for all dynamics. It is therefore important not only to recall the statements of the laws but also to understand them. In exams, candidates often fall down when applying these laws to a variety of situations.

Here is a statement of **Newton's first law** of motion:

> **An object will remain at rest or keep travelling at constant velocity unless it is acted on by an external force.**

This law gives an idea of what a force will do. If the *net* force acting on an object is zero, either it remains motionless or its velocity remains the same. Consider the following situations:

- A cup resting on a table — the cup remains motionless because the net force on it is zero. The weight of the cup is equal to the normal contact force from the table.
- A car travelling at a constant speed on a straight section of a level road — the velocity of the car does not change. Its speed and direction of travel remain unaltered; hence, the net force on the car must be zero. Vertically, the weight of the car is equal to the normal contact force from the road. Horizontally, the forward force acting on the car is balanced by drag and friction.
- A rock in outer space moving with constant velocity — there are no forces acting on the rock, hence, its velocity is unchanged.

The first law is effectively telling us that an object's velocity will change only when it experiences an external force.

Momentum

Before considering Newton's second law of motion, you need to appreciate the term **linear momentum** (or simply, momentum). The momentum p of an object depends on its mass m and its velocity v.

The momentum of an object is defined as follows:

momentum = mass × velocity

or

$$p = mv$$

The unit of momentum is kg m s^{-1}. Momentum is a *vector* quantity; it has both magnitude and direction. The vector nature of momentum implies that you need to be careful when tackling problems either where objects travel in opposite directions or where the direction of travel of an object changes.

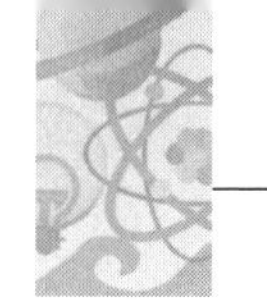

The diagram below shows an object of mass m of 2.0 kg travelling with a speed v of 3.0 m s^{-1}. It collides with a wall and rebounds at the same speed.

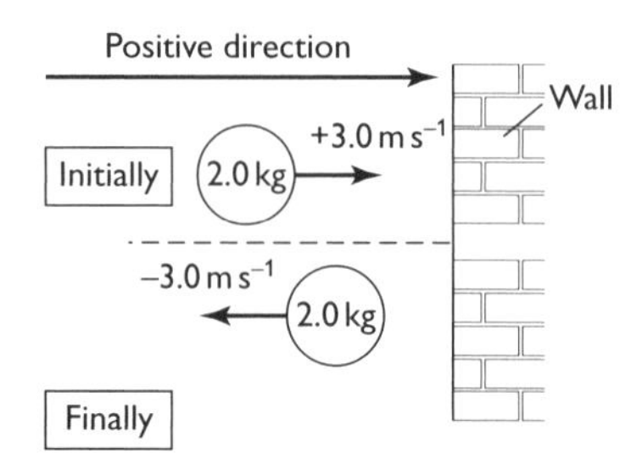

The original direction of travel of the object, left to right, is given a *positive* sign. Therefore:

initial velocity = + 3.0 m s^{-1} — initial momentum = 2.0 × 3.0 = + 6.0 kg m s^{-1}

final velocity = – 3.0 m s^{-1} — final momentum = 2.0 × (– 3.0) = – 6.0 kg m s^{-1}

The magnitude of the momentum is the same. What is the change in the momentum of the object? Many candidates would wrongly state zero. In fact, the magnitude of the change in momentum is twice the initial momentum (see below).

Mathematically:

change in momentum = final momentum – initial momentum

$= (-mv) - (+mv) = -2mv$

$= (-6.0) - (+6.0) = -12$ kg m s^{-1}

Worked example

Calculate the momentum of a 7.5 tonne minibus travelling at 25 m s^{-1} and a 0.80 g wasp travelling in the *opposite* direction at the same speed.

Answer

Minibus: $m = 7.5 \times 10^3$ kg $\quad v = 25$ m s^{-1}

$p = mv = 7.5 \times 10^3 \times 25$

$p = 1.875 \times 10^5$ kg m s$^{-1} \approx 1.9 \times 10^5$ kg m s^{-1}

Wasp: $m = 0.80 \times 10^{-3}$ kg $\quad v = -25$ m s^{-1}

$p = mv = 0.8 \times 10^{-3} \times -25$

$p = -2.0 \times 10^{-2}$ kg m s^{-1}

The momentum of the wasp is negative because it is travelling in the opposite direction to the minibus.

Newton's second law

Too many candidates wrongly state the second law as '$F = ma$', but this equation is just a special case of the second law (see page 18).

Newton's second law of motion relates how quickly (or slowly) the momentum of an object changes according to the force it experiences. A statement of Newton's second law of motion is as follows:

The net force acting on an object is directly proportional to the rate of change of the linear momentum of that object. The net force and the change in momentum are in the same direction.

Therefore:

net force $\propto$ rate of change of momentum

Using SI units (mass in kg, momentum in kg m s^{-1}, time in seconds and force in newtons), the constant of proportionality for the relationship is equal to one. Hence:

net force = rate of change of momentum

Mathematically, this may be written as

$$F = \frac{\Delta p}{\Delta t}$$

where F is the net force acting on the object and Δp is its change of momentum in a time interval of Δt.

You can therefore also write Newton's second law of motion as follows:

The net force acting on an object is *equal* to the rate of change of the linear momentum of that object. The net force and the change in momentum are in the same direction.

Worked example

A 1200 kg car is travelling on a straight road at a speed of 32 m s^{-1}. The brakes are applied for duration of 1.5 s. The final speed of the car is 10 m s^{-1}. Calculate the magnitude of the average force provided by the brakes of the car.

Answer

force = rate of change of momentum

change in momentum $\Delta p = (1200 \times 10) - (1200 \times 32) = -2.64 \times 10^4$ kg m s^{-1}

$$\text{force} = \frac{\Delta p}{\Delta t} = \frac{-2.64 \times 10^4}{1.5}$$

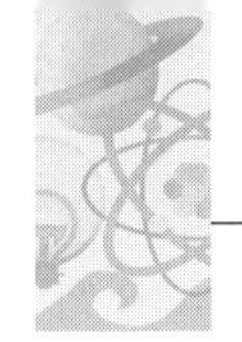

force = -1.76×10^4 N $\approx -1.8 \times 10^4$ N
The magnitude of the force is equal to 1.8×10^4 N.

n Note that the minus sign is omitted because the magnitude of the force is required. The minus sign implies that the braking force is in the opposite direction to car's initial momentum.

A special case of Newton's second law

The equation $F = \frac{\Delta p}{\Delta t}$ is one of the most versatile equations in mechanics. It can be applied to a range of problems, such as predicting the motion of planets in our solar system and modelling how gas molecules exert pressure. However, in some applications, where the mass of the object remains constant, it is easier to use a simplified version of this equation.

The diagram below shows an object of mass m travelling with an initial velocity u. A force F is applied for a time t on the object. After this time, the final velocity of the object becomes v.

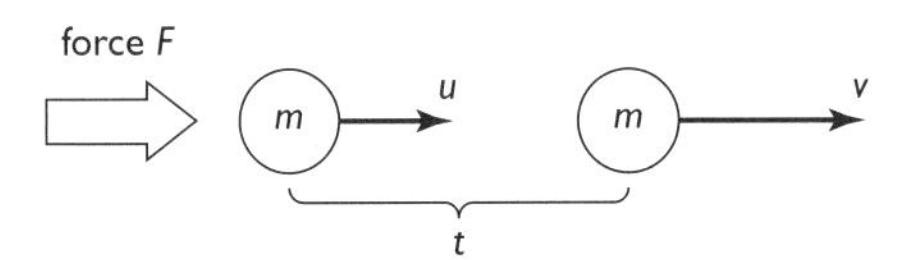

According to Newton's second law of motion, the accelerating force on the object is given by:

$$F = \frac{\Delta p}{\Delta t} = \frac{mv - mu}{t}$$

The mass of the object is constant, so:

$$F = m\left(\frac{v - u}{t}\right)$$

The term in the brackets is the 'rate of change of velocity', or simply, the acceleration, a, of the object. Hence:

$$\boldsymbol{F = ma}$$

You should be familiar with this equation from Unit G481 (Mechanics). Remember that this equation can be used *only* when the mass of the object does not change.

Newton's third law

This is one of the most misunderstood and misquoted laws at advanced level. The problem stems from the popular statement 'action and reaction are equal and opposite'. This statement is not recommended as a formal statement of this law because there are many important facts missing from it. Furthermore, the term **reaction** can mean either a *contact force* between two objects or one of a pair of forces.

When stating Newton's third law, the following points are worth noting about the forces acting on two interacting objects:

- Both forces are of the same type (e.g. electrical, gravitational, magnetic).
- Both forces have the same magnitude.
- The forces are in opposite directions.
- Both forces cannot act on a *single* object.

A clearer statement of Newton's third law is given below.

> **When two objects interact, the forces they exert on each other are equal in magnitude and opposite in direction.**

The diagram below gives some examples.

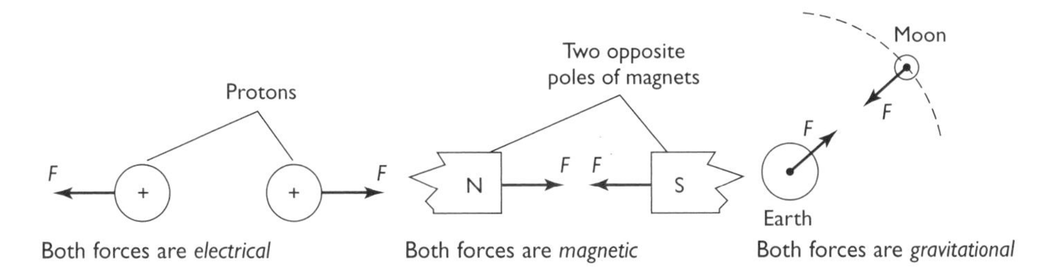

It is worth illustrating this law with a familiar example of a fast-moving car crashing into a safety barrier. The car and the barrier are the two interacting objects. What is the type of the force between the car and the barrier? The force is electrical in origin. It comes about when the objects get close on the atomic scale. The electron clouds associated with the atoms of the two objects repel each other. At all times, the force on the car and force on the barrier have the same magnitude, but of course in opposite directions — this is illustrated in the diagram below.

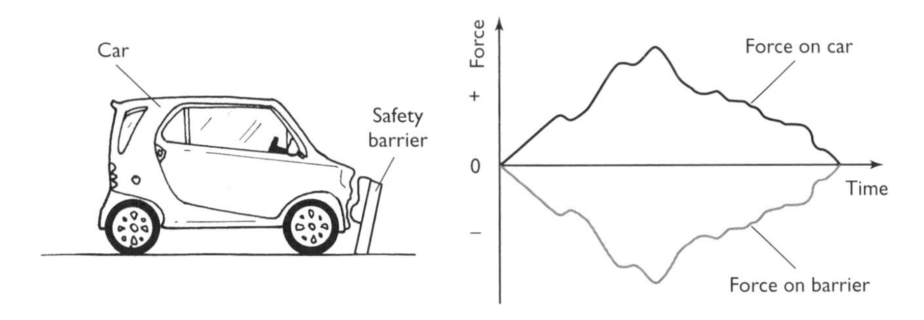

Impulse of a force

According to Newton's second law:

$$F = \frac{\Delta p}{\Delta t}$$

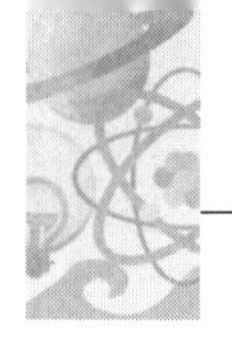

Rearranging this equation gives:

$F \times \Delta t = \Delta p$

The product of force F and the time Δt is known as the **impulse** of the force. The impulse of a force as defined by the following word equation:

impulse = force × time

or

impulse = change in momentum

The quantity impulse can either be measured in N s or kg m s^{-1}.

The impulse of a force is important when the force acting on an object is not constant. Consider the force against time graphs.

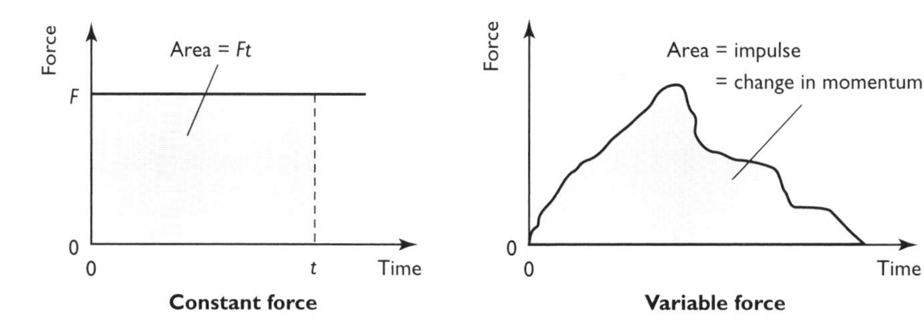

For the graph on the left-hand side, the force is constant. The area under the graph is equal to Ft or the impulse of the force. Moreover, this area is also equal to the change in the momentum of the object. The same is true for the graph on the right-hand side. Therefore:

area under a force against time graph = impulse = change in momentum

Worked example 1

A 900 kg car is travelling at 12 m s^{-1}. A constant force of 1800 N acts on the car for 10 s. Determine the final velocity of the car.

Answer

impulse = force × time

= 1800 × 10 = 1.8×10^4 N s

impulse = change of momentum

Hence, the increase in the momentum of car must be equal to 1.8×10^4 kg m s^{-1}.

final momentum = initial momentum + 1.8×10^4

$900 \times v = (900 \times 12) + 1.8 \times 10^4$

$v = \frac{2.88 \times 10^4}{900} = 32$ m s^{-1}

The final velocity of the car is 32 m s^{-1}.

Note: you could have solved this problem without knowing about impulse. Using $F = ma$ gives an acceleration of 2.0 m s^{-2}. For a 10 s period, the increase in the velocity of the car will be 20 m s^{-1}. The final velocity is the sum of the initial velocity and the increase in the velocity, that is, 32 m s^{-1}. In this question, you have the choice between two methods. However, worked example 2 below cannot be tackled using $F = ma$ and the equations of motion.

Worked example 2

A 46 g golf ball resting on the ground is struck with a club. The graph below shows the variation of the force acting on the ball against time. Calculate the final velocity of the ball.

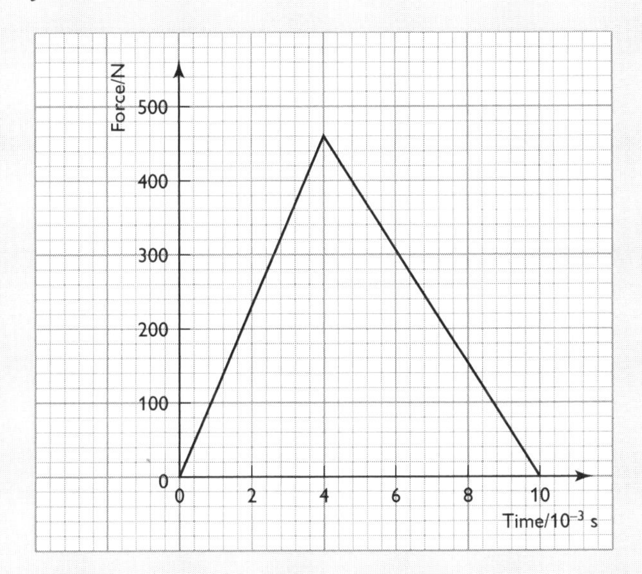

Answer

The area under the force against time graph is equal to the impulse or change in momentum of the ball.

area = $\frac{1}{2} \times \text{base} \times \text{height}$

change in momentum of ball = $\frac{1}{2} \times 10 \times 10^{-3} \times 460 = 2.3$ N s (or kg m s^{-1})

The ball is initially at rest; its initial momentum is zero. Therefore:

final momentum = 2.3 kg m s^{-1}

$0.046v = 2.3$

$v = \frac{2.3}{0.046} = 50$ m s^{-1}

The final velocity of the golf ball is 50 m s^{-1}.

Note: there is no alternative method for determining the final velocity of the ball. You have to use the idea of impulse.

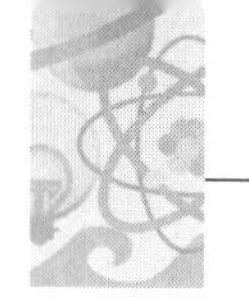

Collisions

Conservation of momentum

What happens when two or more objects interact or collide? The amazing fact is that the total momentum of the objects, in any specified direction, remains the same. We have to consider objects that form a *closed system*; that is, no external forces act during the interaction of the objects. A statement of the principle of conservation of momentum is as follows:

In a closed system, the total momentum in a specified direction remains constant.

There is an alternative to the above statement.

In a closed system, the total momentum of the objects before an interaction is equal to the total momentum of the objects after the interaction.

Whenever solving problems on momentum, you must not forget the vector nature of momentum. Consider a situation where two objects A and B travelling in opposite directions collide head on, and then stick together after the collision. The diagram below shows the situation before and after the collision.

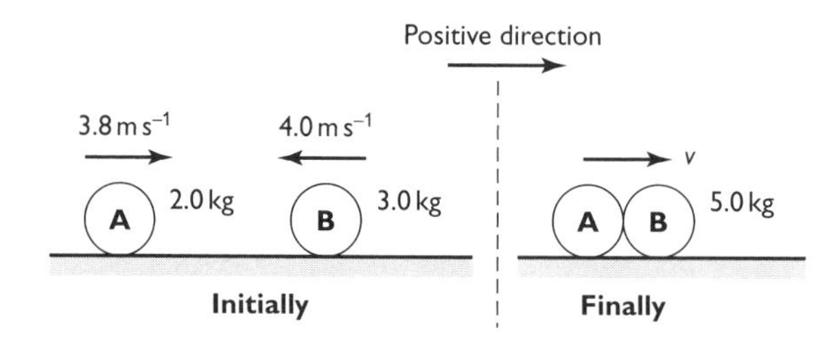

The positive direction is taken to be from left to right. We can determine the common velocity v of the objects by using the principle of conservation of momentum.

total initial momentum = total final momentum

$(2.0 \times 3.8) + (3.0 \times -4.0) = 5.0 \times v$ (The total mass after the collision is 5.0 kg.)

Therefore:

$-4.4 = 5.0v$

$v = \frac{-4.4}{5.0} = -0.88 \text{ m s}^{-1}$

The speed of the combined objects is 0.88 m s^{-1}. The negative sign means that the objects travel towards the *left* after the collision.

Worked example

Two cars **A** and **B** travelling in the same direction collide. The momentum against time graphs for these two cars are shown below.

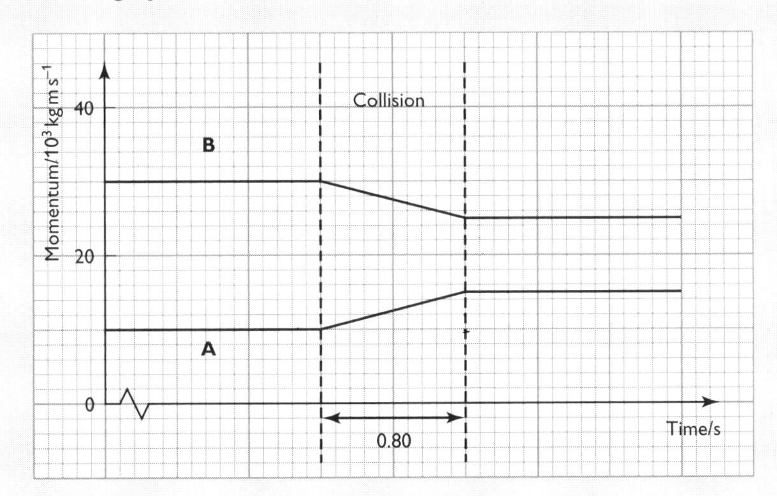

Use the graph to:

(a) show that momentum is conserved

(b) determine the force acting on each car during the collision

Answers

(a) total initial momentum = $(30 + 10) \times 10^3 = 4.0 \times 10^4$ kg m s^{-1}

total final momentum = $(25 + 15) \times 10^3 = 4.0 \times 10^4$ kg m s^{-1}

The total initial momentum is equal to the total final momentum; hence momentum is conserved.

(b) During the collision, the momentum of each car changes. The force on each car can be determined from Newton's second law. That is:

force = rate of change of momentum

For car **A**: force $= \frac{\Delta p}{\Delta t} = \frac{(15-10)\times 10^3}{0.80} = 6.25 \times 10^3$ N

For car **B**: force $= \frac{\Delta p}{\Delta t} = \frac{(25-30)\times 10^3}{0.80} = -6.25 \times 10^3$ N

The magnitude of the force experienced by each car is about 6.3 kN.

Note: each car experiences an equal but opposite force, as expected from Newton's third law.

Why is momentum conserved?

Why is it that in all interactions momentum is conserved? This can be explained using Newtonian laws. Consider two objects **A** and **B** colliding. During the collision, the force each object exerts on the other is the same but opposite in directions. That is:

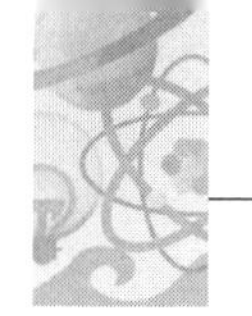

force exerted on **A** by **B** = –force exerted on **B** by **A**

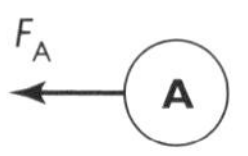

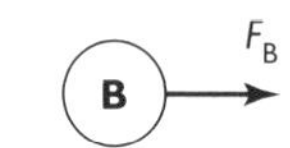

The minus sign simply emphasises that the two forces are in the opposite direction. Using Newton's second law gives:

$$\frac{\Delta p_A}{\Delta t} = -\frac{\Delta p_B}{\Delta t}$$

where Δp_A is the change of momentum of **A**, Δp_B is the change of momentum of **B** and Δt is the time of the collision (which must be the same for each object). The time Δt can be cancelled on either side of the equation above and this leaves:

$$\Delta p_A = -\Delta p_B$$

This means that if **A** gains momentum, **B** must lose an equal amount of momentum (or vice versa). In other words, the total momentum of the two objects remains unchanged: momentum is conserved. The principle of conservation of momentum is therefore a logical conclusion from Newton's second and third laws.

Perfectly elastic and inelastic collisions

In all collisions or interactions, the following quantities are always conserved:

- total energy
- total momentum

The principle of conservation of energy cannot be broken, so some of the energy can be transformed to other forms such as heat and sound. There are two types of collisions:

- perfectly elastic collision
- inelastic collision

The definitions for these types of collisions are stated below:

In a perfectly elastic collision, the total kinetic energy of the system remains constant. Momentum and total energy are also conserved.

In an inelastic collision, the total kinetic energy is not conserved. Some of the kinetic energy is transformed into heat, sound etc. Momentum and total energy are both conserved.

Worked example

The diagram opposite shows the 'before' and 'after' situations for two colliding toy cars.

(a) Determine the unknown speed *v* of one of the toy cars.

(b) Deduce the type of collision between the toy cars.

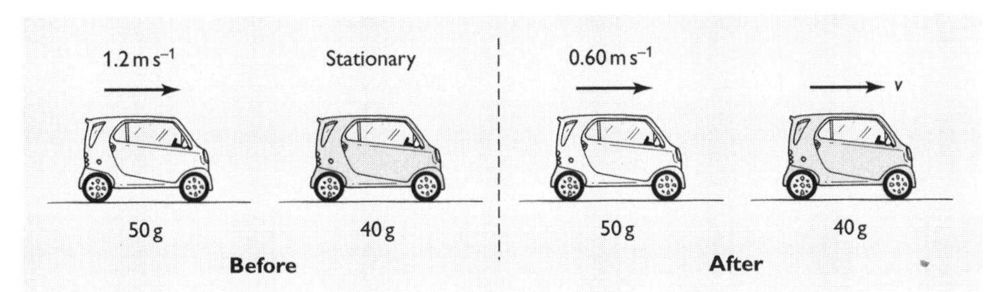

Answer

(a) Using the principle of conservation of momentum, we have total initial momentum = total final momentum

$(0.050 \times 1.2) + (0.040 \times 0) = (0.050 \times 0.60) + 0.040v$

$0.060 = 0.030 + 0.04v$

$0.060 - 0.030 = 0.040v$

$v = \frac{0.030}{0.040} = 0.75 \text{ m s}^{-1}$

The toy car, initially at rest, has a final velocity of 0.75 m s^{-1} to the *right* immediately after the collision.

(b) kinetic energy $E_k = \frac{1}{2}mv^2$

total initial kinetic energy = $\frac{1}{2} \times 0.050 \times 1.2^2 + 0 = 0.036$ J

total final kinetic energy = $\frac{1}{2} \times 0.050 \times 0.60^2 + \frac{1}{2} \times 0.040 \times 0.75^2 \approx 0.020$ J

There is a loss of kinetic energy, so the collision must be inelastic.

Zero momentum?

What happens when a bullet is fired from a gun? The initial momentum of the system (the gun and bullet) is zero. According to the principle of conservation of momentum, the final momentum must also be zero. The bullet flies off in one direction and the gun must recoil in the opposite direction with the same momentum.

n Note: momentum is a vector. If the momentum of the bullet is positive, then the momentum of the gun must be negative.

Here is another situation of zero momentum. A person jumps off a wall and falls towards the Earth. The velocity of the person increases as it falls. The person gains momentum. Does this mean that the principle of conservation of momentum is violated? The simple answer is no. The diagram on page 26 shows that if at a given instant the person has a momentum of +400 kg m s^{-1}, the Earth must have a momentum of −400 kg m s^{-1}. These two numbers add up to zero.

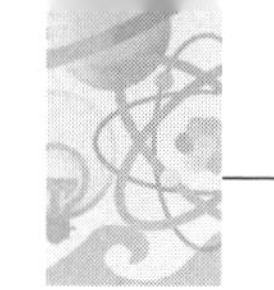

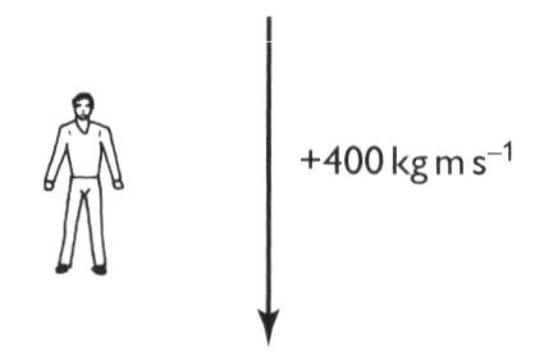

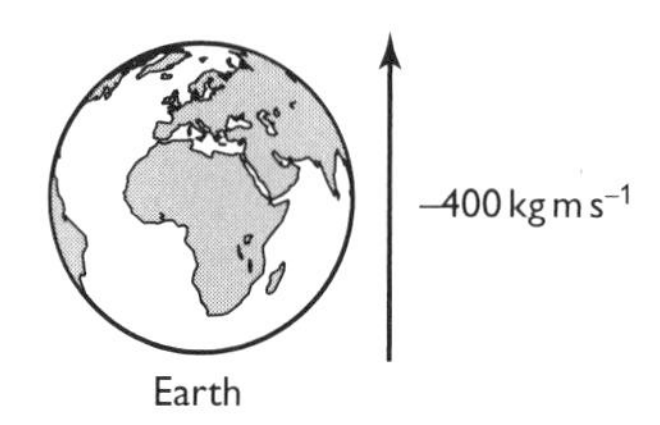

The total momentum is zero

Worked example

An astronaut is 'floating' in a space capsule and has no motion relative to the capsule. The astronaut has total mass 90 kg and is holding a 2.0 kg spanner. The astronaut throws the spanner away from him at a speed of 8.0 m s^{-1}. Calculate the speed of the astronaut and describe his subsequent motion.

Answer

The initial momentum of the astronaut and the spanner is zero. Momentum is always conserved, so the astronaut will travel in a direction *opposite* to that of the spanner. Both the spanner and the astronaut will have momentum of the same magnitude. Therefore, the speed v of the astronaut can be determined as follows:

$$90 \times v = 2.0 \times 8.0$$

$$v = \frac{16}{90} \approx 0.18 \text{ m s}^{-1}$$

The astronaut travels in a direction opposite to that of the spanner at a constant speed of about 18 cm s^{-1}.

Circular motion

Angles in radians

The diagram opposite shows an object travelling in a circular path. The angle θ it travels through is known as its **angular displacement**.

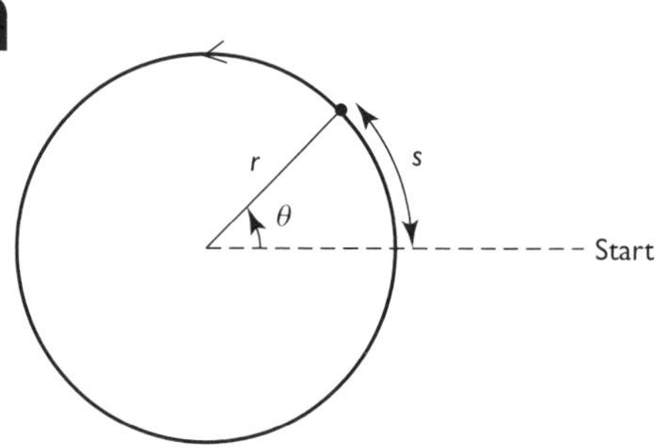

When analysing motion of objects in a circle or oscillatory motion (see simple harmonic motion on page 37), it is convenient to have the angle in radians (abbreviated to rad). The angle θ in radians is defined by the following word equation:

$$\theta = \frac{\textbf{arc length}}{\textbf{radius}}$$

Or simply

$$\theta = \frac{s}{r}$$

The angle is 1 radian when the arc length s is equal to the radius r of the circle. For an angle of 360°, the arc length is equal to the circumference $2\pi r$ of the circle. Therefore, an angle of 360° is equivalent to 2π radians. Similarly, π rad = 180°.

Motion in a circle

The diagram below shows an object (e.g. a conker at the end of a string, a planet orbiting a star, a car turning a corner) travelling at a constant speed v in a circle of radius r.

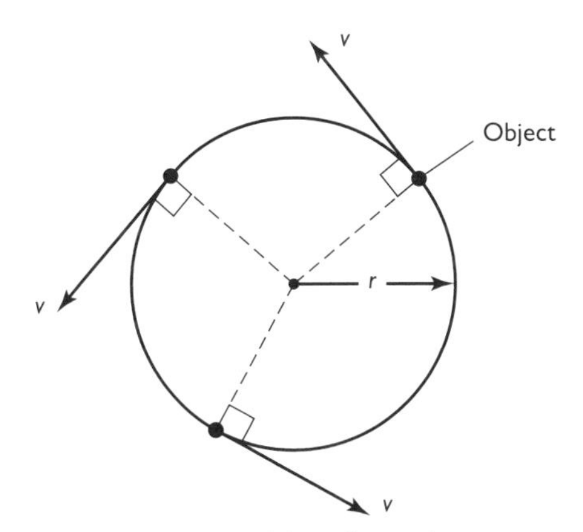

The direction of the velocity changes

It is vital to appreciate that the *velocity* of any object moving in a circle changes because its *direction* of travel changes. Since 'acceleration = rate of change of velocity', the implication is that the object has an acceleration (see page 29).

The speed v of the object can be determined from the circumference of the circle and the period T.

Note: the period is the time taken to complete one revolution.

$$\text{speed} = \frac{\text{distance}}{\text{time}}$$

$$v = \frac{2\pi r}{T}$$

Worked example

A metal disc of diameter 12 cm is spun at a rate of 480 revolutions per minute.

(a) Describe how the speed of a point on the disc depends on its distance from the centre.

(b) Calculate the maximum speed of a point on this spinning disc.

Answer

(a) The speed v of any point on the disc is given by the equation $v = \frac{2\pi r}{T}$.

The period T is the same for all points on the disc, therefore the speed v is directly proportional to the radius r.

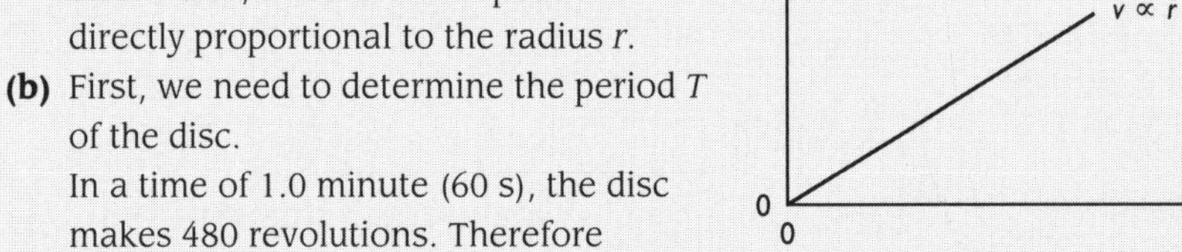

(b) First, we need to determine the period T of the disc.

In a time of 1.0 minute (60 s), the disc makes 480 revolutions. Therefore

$$T = \frac{60}{480} = 0.125\ \text{s}$$

A particle on the rim of the disc will have maximum speed v; the radius r of the circle is 6.0 cm. Therefore:

$$v = \frac{2\pi r}{T} = \frac{2\pi \times 6.0 \times 10^{-2}}{0.125}$$

$$v = 3.02\ \text{m s}^{-1} \approx 3.0\ \text{m s}^{-1}$$

n Note: it is important to work in SI units throughout. Many candidates incorrectly substitute the value of the diameter rather than the radius when calculating the speed and overlook the conversion from minutes to seconds.

Centripetal acceleration and force

The diagram on page 29 shows an object travelling at constant speed v. The object must have acceleration because its velocity is changing — its direction of travel is changing.

n Note: the magnitude of the velocity does not change, so the acceleration must always be at right angles to the velocity.

The acceleration a of the object is always directed towards the centre of the circle. This acceleration is known as the **centripetal acceleration** of the object. Since $F = ma$, the *net* force F on the object must also be pointing towards the centre. This net force is called the **centripetal force**. The net force providing the circular motion could be a single fundamental force, such as gravitational force, or the resultant of several forces. Look at the examples in the table opposite.

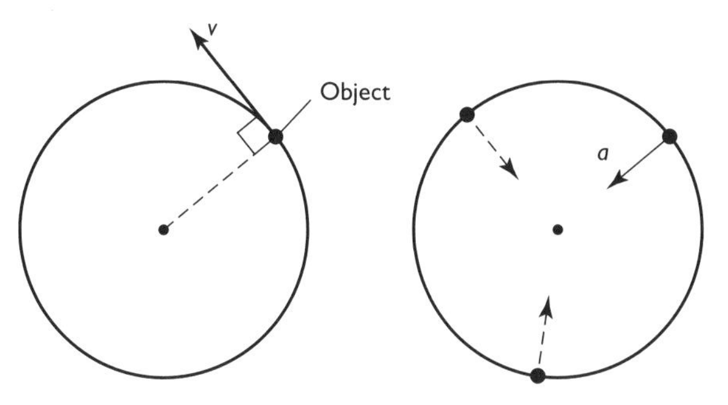

The centripetal acceleration is at 90° to the velocity

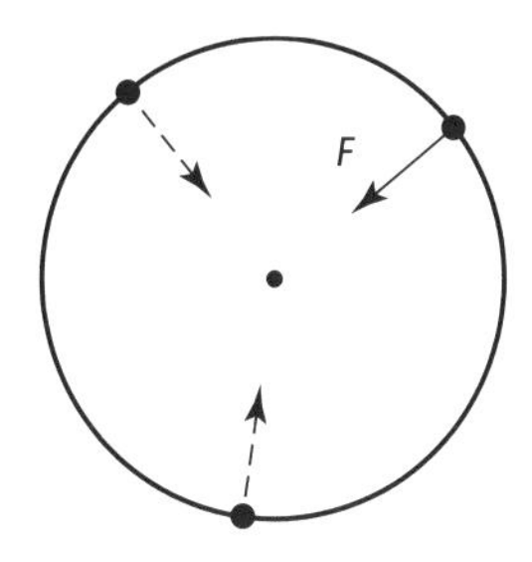

The centripetal force and acceleration are in the same direction

Situation	What provides the force?
A planet orbiting the Sun.	The *gravitational* force between the planet and the Sun.
A rubber bung attached to a string being whirled in a horizontal circle.	The *tension* in the string
An electron 'orbiting' the nucleus.	The *electrical* force between the positive nucleus and the negative electron.
A car going round a bend.	*Friction* between the tyres and the road.
A pilot in a jet doing a vertical loop in the sky.	The *weight* of the pilot and the *contact force* between the pilot and the seat.

The centripetal acceleration a and the centripetal force F are given by the following equations:

$$\boldsymbol{a} = \frac{\boldsymbol{v}^2}{\boldsymbol{r}} \qquad \text{and} \qquad \boldsymbol{F} = \boldsymbol{ma} = \frac{\boldsymbol{mv}^2}{\boldsymbol{r}}$$

where v is the speed of the object describing a circle of radius r.

Worked example 1

Calculate the orbital speed and centripetal acceleration of the Earth as it orbits round the Sun at a mean distance of 1.5×10^{11} m. (1 year = 3.16×10^7 s)

Answer

The orbital speed v can be determined as follows:

$$v = \frac{2\pi r}{T} = \frac{2\pi \times 1.5 \times 10^{11}}{3.16 \times 10^7}$$

$v = 2.983 \times 10^4$ m s^{-1} (30 km s^{-1})

The orbital speed of the Earth is about 30 km s^{-1}.

The centripetal acceleration a is given by the equation $a = \frac{v^2}{r}$. Therefore:

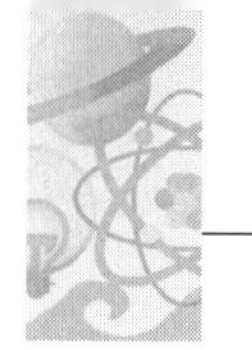

$$a = \frac{v^2}{r} = \frac{(2.983 \times 10^4)^2}{1.5 \times 10^{11}} \approx 5.9 \times 10^{-3} \text{ m s}^{-2}$$

The centripetal acceleration of the Earth (towards the centre of the Sun) is about 5.9×10^{-3} m s^{-2}.

Worked example 2

'Synopticity' and 'stretch and challenge'

A rubber bung of mass 50 g is attached to a length of string. The free end of the string is fixed to a motor and the bung is whirled in a horizontal circle of radius 30 cm. The string has a cross-sectional area of 4.0×10^{-8} m^2 and breaking stress 5.0×10^9 Pa. The speed of the bung is increased gradually until the string breaks. Estimate the maximum speed of the bung.

Answer

The maximum force F provided by the tension in the string can be calculated from its breaking stress.

$$\text{breaking stress} = \frac{\text{maximum force}}{\text{cross-sectional area}}$$

Therefore, $F = 5.0 \times 10^9 \times 4.0 \times 10^{-8} = 200$ N

$$F = \frac{mv^2}{r}$$

$$v = \sqrt{\frac{Fr}{m}} = \sqrt{\frac{200 \times 0.30}{0.050}}$$

$v \approx 35$ m s^{-1}

The maximum speed of the rubber bung is about 35 m s^{-1}.

n Note: the calculation is an estimate because it is assumed that neither the length nor the thickness of the string changes under increasing tension.

Gravitational fields

Existence of gravitational fields

All objects have mass. The mass of the object creates a gravitational field in the space around the object. How can it be known that a gravitational field exists at a particular point in space? All that needs to be done is to place a small mass at that point and observe whether it experiences a force.

The gravitational field pattern can be mapped out using the idea of *field lines*. The *direction* of the field is the direction in which a small mass would move. Here are some examples:

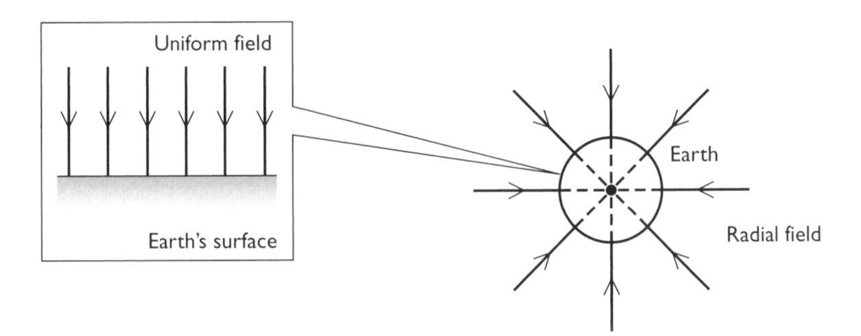

On the surface of the Earth, the gravitational field lines are equally spaced and parallel (because the Earth has a large radius); the gravitational field is *uniform*. Further away from the Earth, the field pattern becomes *radial*. The field lines stop at the surface of the Earth, but appear to converge at the Earth's centre. This implies that we can model the Earth as being a *point mass*.

Newton's law of gravitation

Gravitational force is an attractive force. It exists between all particles that have mass. According to Newton's third law, two particles (also referred to as point masses) close to each other will interact to produce equal and opposite forces.

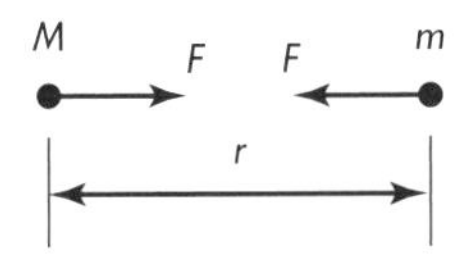

Newton's law of gravitation states that:

> **Two point masses will attract each other with a force that is directly proportional to the product of their masses and inversely proportional to the square of their separation.**

Mathematically:

force ∝ product of masses $\qquad F \propto Mm$

$$\text{force} \propto \frac{1}{\text{separation}^2} \qquad F \propto \frac{1}{r^2}$$

Therefore:

$$F \propto \frac{Mm}{r^2}$$

Using the gravitational constant G (6.67×10^{-11} N m^2 kg^{-2}), the following equation can be written:

$$\boldsymbol{F = -\frac{GMm}{r^2}}$$

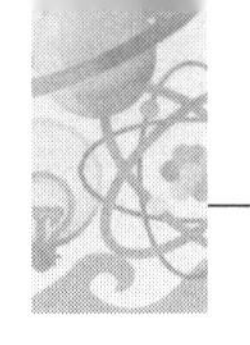

The minus sign in the equation indicates the gravitational force is attractive. It is also worth noting that the force decreases with the square of the separation — double the separation and the force decreases by a factor of 4, treble the separation and the force decreases by a factor of 9 etc.

Note: you can use the above equation for spherical objects as long as you measure the separation between the centres. Hence, the Earth can be modelled as though its entire mass was concentrated at its centre.

Worked example

A 400 kg satellite orbits the Earth at a distance of 1.5×10^4 km from the centre of the Earth. The mass of the Earth is 6.0×10^{24} kg. Calculate:

(a) the gravitational force acting on the satellite

(b) the acceleration and orbital speed of the satellite

Answer

(a) The Earth is considered as though its mass is concentrated at its centre. The magnitude of the force F acting on the satellite is:

$$F = \frac{GMm}{r^2} = \frac{6.67 \times 10^{-11} \times 6.0 \times 10^{24} \times 400}{(1.5 \times 10^4 \times 10^3)^2}$$

$$F = 711.5 \text{ N} \approx 710 \text{ N}$$

(b) The gravitational force on the satellite provides the centripetal force. The acceleration a of the satellite is:

$$a = \frac{F}{m} = \frac{711.5}{400} = 1.78 \text{ m s}^{-2} \approx 1.8 \text{ m s}^{-2}$$

The orbital speed v is given by the equation $a = \frac{v^2}{r}$, therefore:

$$v = \sqrt{ar} = \sqrt{1.78 \times 1.5 \times 10^7}$$

$$v = 5.17 \times 10^3 \text{ m s}^{-1} \text{ (5.2 km s}^{-1}\text{)}$$

Note: you need to use the mass of the satellite, and not that of the Earth, when determining the acceleration of the satellite in its circular orbit round the Earth.

Gravitational field strength

The gravitational field strength g at a point in space is defined as follows:

The gravitational field strength at a point in space is the gravitational force experienced per unit mass on a small object placed at that point.

This is written mathematically as:

$$g = \frac{F}{m}$$

The unit for gravitational field strength is N kg^{-1}.

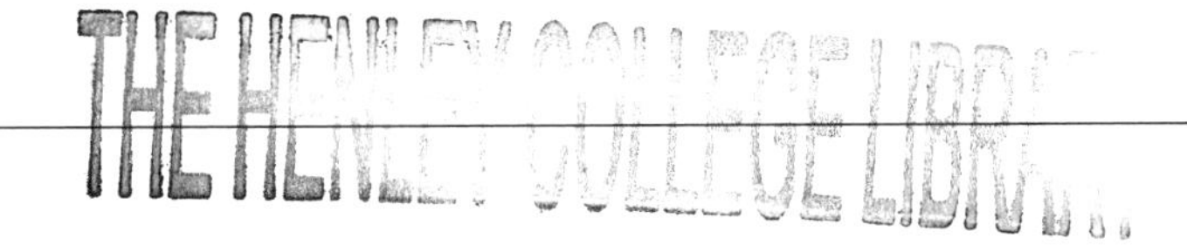

Since the ratio of force to mass is also the acceleration of free fall of an object, it can also be deduced that the magnitude of the acceleration a of free fall of an object is equal to the gravitational field strength g:

$$\boldsymbol{a = g}$$

This is not surprising. The acceleration of free fall on the Earth's surface is 9.81 m s^{-2}, so the gravitational field strength must be 9.81 N kg^{-1}.

Note: you have already met a version of the equation $g = \frac{F}{m}$ in Unit G481. The gravitational force F acting on an object on the Earth is known as its weight, and weight = mg.

What is the magnitude of the gravitational field strength for a spherical object (e.g. the Earth) or a point mass?

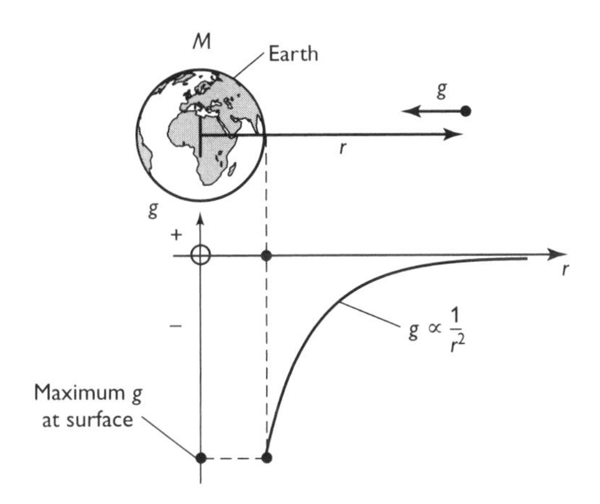

$$g = \frac{\text{force}}{\text{mass}}$$

$$g = \left(-\frac{GMm}{r^2}\right) \div m$$

$$\boldsymbol{g = -\frac{GM}{r^2}}$$

The gravitational field strength due to an object of mass M also obeys an inverse square law with distance r. The negative sign implies that gravity provides an attractive field. Remember that all distances are measured from the centre of the gravitating mass.

Worked example

Saturn is a giant gaseous planet with surface gravitational field strength at its surface of 10.4 N kg^{-1}. The radius of the planet is 6.04×10^4 km. Determine the mass and mean density of the planet.

Answer

Saturn is modelled as though its mass is concentrated at its centre.

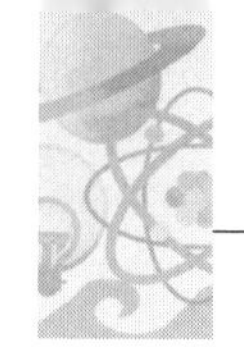

$$g = \frac{GM}{r^2}$$ (The negative sign is omitted.)

$$M = \frac{gr^2}{G} = \frac{10.4 \times (6.04 \times 10^7)^2}{6.67 \times 10^{-11}}$$

$$M = 5.69 \times 10^{26}\ \text{kg}$$

The density ρ of the planet can be determined by dividing its mass by its volume. That is:

$$\rho = \frac{m}{V} = \frac{5.69 \times 10^{26}}{\frac{4}{3}\pi \times (6.04 \times 10^7)^3}$$

$$\rho = 616\ \text{kg m}^{-3} \approx 620\ \text{kg m}^{-3}$$

Note: it is worth noting that the density of Saturn is less than the density of water (1000 kg m^{-3}); Saturn is a gaseous planet.

Circular orbits

Here are some examples where gravitational force provides circular motion:

- planets moving round the Sun in our solar system
- the Moon orbiting the Earth
- an artificial satellite orbiting the Earth
- the Sun orbiting the centre of our galaxy (the Milky Way)

The physics of all these examples is the same. Gravity provides the necessary centripetal force for circular motion. The following four equations are all you need to solve problems like the ones above.

$$F = \frac{GMm}{r^2}\ (1) \qquad F = ma\ (2) \qquad a = \frac{v^2}{r}\ (3) \qquad v = \frac{2\pi r}{T}\ (4)$$

Consider a real or artificial satellite of mass m orbiting a planet of mass M. The radius of the circular orbit is r and the satellite has an orbital period T. What is the relationship between the orbital period T and its distance r from the centre of the planet? Using equations (1) and (2) gives:

$$F = ma$$

$$\frac{GMm}{r^2} = ma$$

The mass m cancels on both sides of the equation, leaving:

$$a = \frac{GM}{r^2}$$

This equation should be familiar as we have already equated acceleration of free fall with gravitational field strength (page 33).

Using equation (3) gives:

$$\frac{v^2}{r} = \frac{GM}{r^2}$$

$$v^2 = \frac{GM}{r}$$

Finally, using equation (4) gives:

$$\left(\frac{2\pi r}{T}\right)^2 = \frac{GM}{r}$$

Rearranging this gives the final equation below (note: you do need to know the above derivation for the exam and the equation is given in the *Data, Formulae and Relationships* booklet).

$$\boldsymbol{T^2 = \left(\frac{4\pi^2}{GM}\right) r^3}$$

The most fascinating observation is that for a given gravitating mass M, the square of the period is directly proportional to the cube of the distance. This idea can be applied to the planets in our solar system. This relationship is known as Kepler's third law of planetary motion:

The square of the period T of a planet is directly proportional to the cube of its distance r from the Sun.

Kepler's third law can be written as $\boldsymbol{T^2 \propto r^3}$.

Candidates are usually happy to substitute numbers into equations, but many have serious problems in conveying their understanding of a relationship such as Kepler's third law.

Worked example

The planet Jupiter has numerous moons. Europa and Callisto are two larger moons of Jupiter. Europa orbits at a distance of 6.7×10^5 km from Jupiter's centre and has a period of 3.55 days. Calculate the distance of Callisto from Jupiter's centre given its orbital period is 16.7 days.

Answer

When using Kepler's third law, you can use any system of units as long as you are consistent.

Europa: $T = 3.55$ days $r = 6.7 \times 10^5$ km

Callisto: $T = 16.7$ days $r = ?$

According to Kepler's law, we have $T^2 \propto r^3$. That is

$$\frac{T^2}{r^3} = \text{constant}$$

Therefore:

$$\frac{3.55^2}{(6.7 \times 10^5)^3} = \frac{16.7^2}{r^3}$$

$$r^3 = \frac{16.7^2 \times (6.7 \times 10^5)^3}{3.55^2} = 6.6558 \times 10^{18}$$

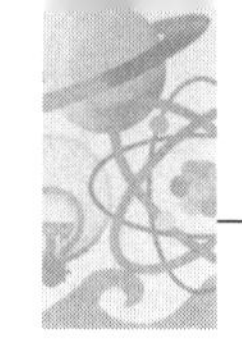

$r = \sqrt[3]{6.6558 \times 10^{18}} \approx 1.88 \times 10^{6}$ km

Callisto travels round Jupiter in a circular orbit of radius 1.88×10^{6} km.

Note: there is an alternative technique. Since $r^3 \propto T^2$, $r \propto T^{2/3}$. The period increases by a factor of $\frac{16.7}{3.55} = 4.7042$, so the radius must increase by a factor of $4.7042^{2/3} = 2.8075$. Therefore, Callisto's distance from the centre of Jupiter must be distance = $(6.7 \times 10^{5}) \times 2.8075 \approx 1.88 \times 10^{6}$ km.

Geostationary satellites

Scientists use satellites for communications, exploration etc. For communications, satellites are particularly useful if they are in a *geostationary* orbit. The period of the satellite is exactly equal to the rotational period of the Earth — 1 day. Therefore, it remains above the same place on the Earth. It is important to appreciate that a geostationary orbit is only possible above the equator.

How far from the Earth's surface is a geostationary satellite? You can determine the period T of a satellite or its distance r from the centre of the Earth using the equation:

$$T^2 = \left(\frac{4\pi^2}{GM}\right) r^3$$

(The mass of the Earth is 6.0×10^{24} kg.)

Therefore:

$$(1 \times 24 \times 3600)^2 = \left(\frac{4\pi^2}{6.67 \times 10^{-11} \times 6.0 \times 10^{24}}\right) \times r^3$$

$$r \approx 4.23 \times 10^{7} \text{ m}$$

The radius R of the Earth is 6400 km. A geostationary satellite orbits at a distance of $6.5R$ from the centre of the Earth. Hence, it orbits at a height of $5.5R$ above the Earth's surface.

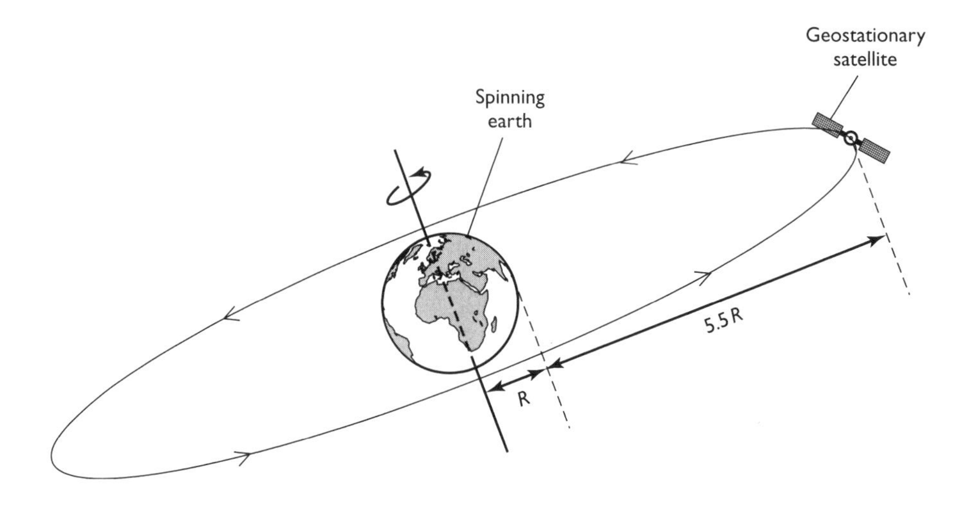

Simple harmonic motion

Defining quantities

When free to move, all oscillators have a to-and-fro motion about some fixed (or equilibrium) point. The oscillator could be a mass hanging vertically from a spring, the atoms in a solid, a bridge etc. It is vital that you learn the following definitions (which may look familiar from your study of waves):

- Displacement x — the distance moved by the oscillator in a specified direction from its equilibrium position. Unit: metres (m).
- Amplitude A — the magnitude of the maximum displacement of the oscillator. Unit: metres (m).
- Period T — the time taken for one complete oscillation. Unit: seconds (s).
- Frequency f — the number of oscillations per unit time. Unit: hertz (Hz).
  The frequency f of an oscillator and its period T are linked by the equation:

  $$f = \frac{1}{T}$$

- Phase difference ϕ — this is the fraction of an oscillation one oscillator leads or lags behind another oscillator. Phase difference can be quoted in radians or in degrees. The first diagram below illustrates a phase difference of π rad (180°) and the second a phase difference of π/2 rad (90°).

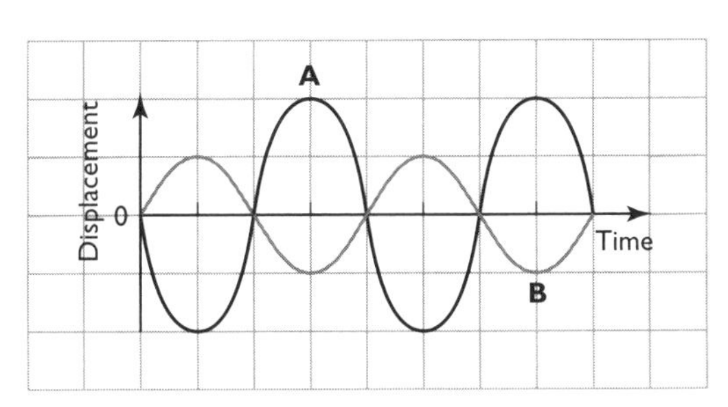

Oscillations **A** and **B** are
180° out of phase

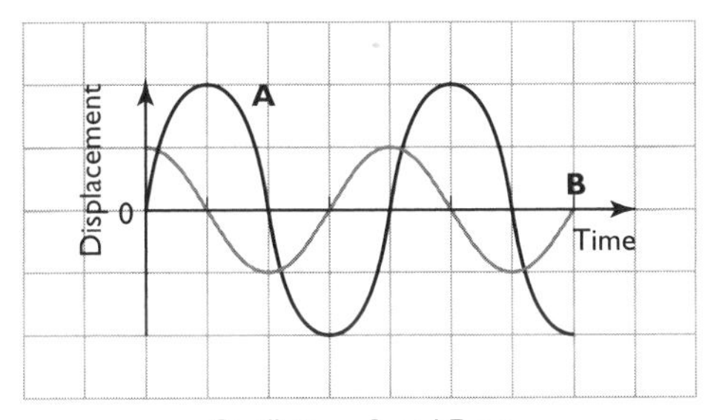

Oscillations **A** and **B** are
90° out of phase

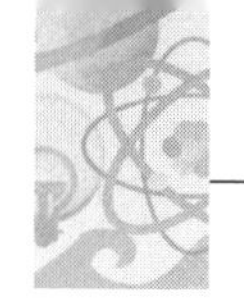

Simple harmonic motion

The simplest motion that can be modelled for an oscillator is known as simple harmonic motion. The statement below defines an oscillator executing simple harmonic motion.

A body executes simple harmonic motion when its acceleration is directly proportional to its displacement from its equilibrium position, and is directed towards the equilibrium position.

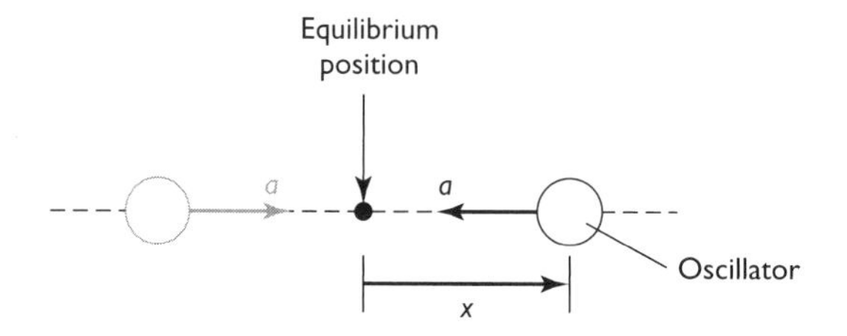

For an oscillator with this type of motion:

acceleration ∝ –displacement

or

$a \propto -x$

The minus sign is a mathematical way of stating that the direction of the acceleration is always towards the equilibrium position. The constant of proportionality is equal to $2\pi f$, where f is the frequency of the oscillator. Therefore, the equation for acceleration is:

$$a = -(2\pi f)^2 x$$

The quantity $2\pi f$ is known as the **angular frequency** ω of the oscillator. Since $f = 1/T$, we can also have $\omega = 2\pi/T$.

Worked example

A mass is attached to the end of a spring. The mass is pulled down and released. The mass performs simple harmonic motion. In a time of 10 s, the mass performs 25 oscillations. The amplitude of the motion is 8.0 cm.

(a) Calculate the maximum acceleration a_{max} of the mass.

(b) Sketch a graph of acceleration against displacement for this oscillator.

Answer

(a) The maximum acceleration a_{max} occurs when the mass has maximum displacement.

$a_{max} = (2\pi f)^2 x$

We need to first find f.

$$f = \frac{\text{number of oscillations}}{\text{time}} = \frac{25}{10} = 2.5 \text{ Hz}$$

$a_{max} = (2\pi \times 2.5)^2 \times 0.08$
$= 19.7 \text{ m s}^{-2} \approx 20 \text{ m s}^{-2}$

(b) For a simple harmonic oscillator, $a \propto -x$.
Hence a graph of a against x will be a straight line through the origin. The slope of the line will be negative (because of the minus sign). The magnitude of the gradient of the line is equal to $(2\pi f)^2$ — about 250 rad² s⁻².

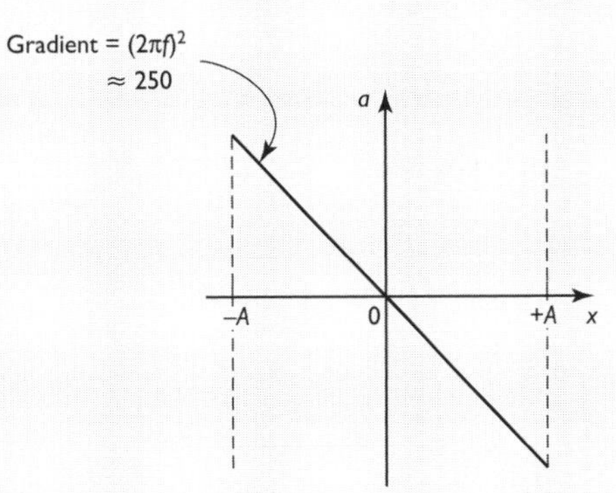

Displacement, velocity and acceleration

The displacement x of a simple harmonic oscillator is given by the following *sinusoidal* equations, where A is the amplitude:

$\boldsymbol{x = A\cos(2\pi ft)}$ and $\boldsymbol{x = A\sin(2\pi ft)}$

It is vital that your calculator is in 'radian mode' when using these equations. You use the cosine equation when an oscillator starts ($t = 0$) from its maximum displacement ($x = A$) and the sine equation when the oscillator starts from its equilibrium position ($x = 0$).

The set of graphs below show the variation of displacement (x), velocity (v) and acceleration (a) of an oscillator with time.

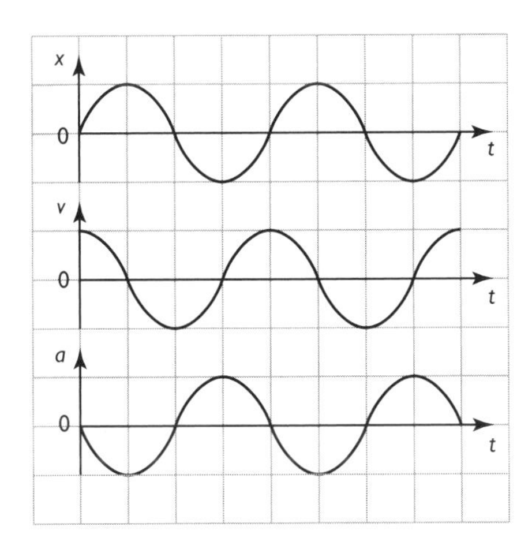

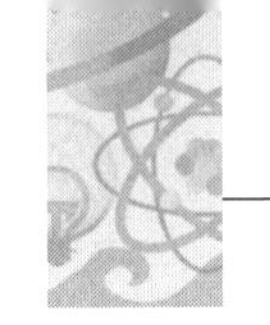

- The velocity is equal to the gradient of the displacement against time graph. The acceleration is equal to the gradient of the velocity against time graph. There is a phase difference of $\pi/2$ rad (90°) between velocity and displacement.
- The phase difference between the acceleration and displacement is 180°, that is $a \propto -x$.

The maximum speed of an oscillator occurs when it travels through the equilibrium position ($x = 0$). The maximum speed v_{max} of the oscillator is given by the equation below:

$$\mathbf{v_{max} = (2\pi f)A}$$

where A is the amplitude of the oscillator and f is its frequency.

Energy of a simple harmonic oscillator

An oscillator has maximum speed as it travels through the equilibrium position and is momentarily at rest when its displacement equals the amplitude. The kinetic energy of the oscillator is zero at the extremes of its motion but the oscillator has maximum potential energy at these points. There is interchange between potential energy and kinetic energy as a pendulum swings to and fro or a spring-mass system bobs up and down.

The graph below shows the variation of energy of the oscillator with displacement.

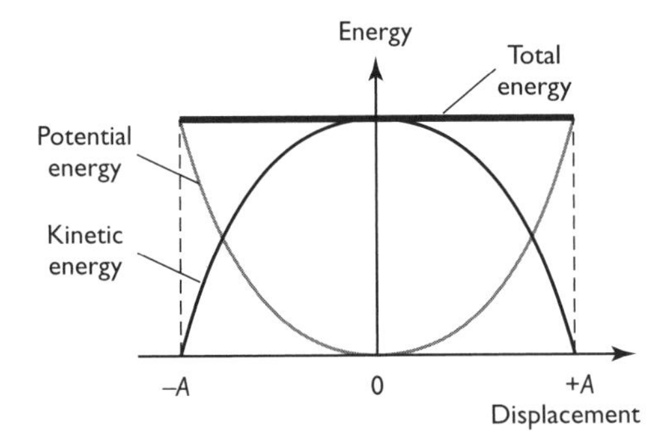

Worked example

A pendulum bob of mass 60 g executes simple harmonic motion with amplitude of 12 cm. The period of the pendulum is 0.52 s. Calculate:

(a) the maximum speed v_{max} of the pendulum bob

(b) the maximum change in its gravitational potential energy E_p

Answer

(a) $v_{max} = (2\pi f)A$

Since $f = 1/T$, we also have $v_{max} = \left(\frac{2\pi}{T}\right)A$.

Therefore:

$$v_{max} = \left(\frac{2\pi}{0.52}\right) \times 0.12 = 1.45 \text{ m s}^{-1} \approx 1.5 \text{ m s}^{-1}$$

(b) The maximum value of E_p must be equal to the maximum kinetic energy E_k of the pendulum bob, hence:

$$E_p = E_k = \frac{1}{2}mv^2 = \frac{1}{2} \times 0.060 \times 1.45^2 \approx 6.3 \times 10^{-2}\ \text{J}$$

Damped motion

Damped motion is the result of friction acting on the oscillator. The displacement x against time t graph is no longer a pure sinusoidal graph. The greater the degree of damping, the greater is rate of energy loss from the oscillator and hence the faster its amplitude decays.

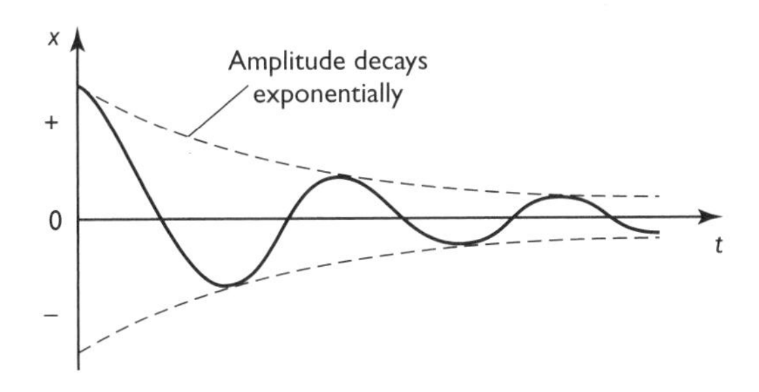

Resonance

Forced oscillations occur when an object is forced to vibrate at the frequency of an external source. A bus engine, for example, makes parts such as the seats of the bus vibrate at the frequency of the engine. When the frequency of the engine matches the natural frequency of the seat, the seat *resonates* — the amplitude of its vibrations becomes considerably larger.

The graph of the amplitude A of a forced oscillator against the forcing frequency f has a characteristic shape — see below.

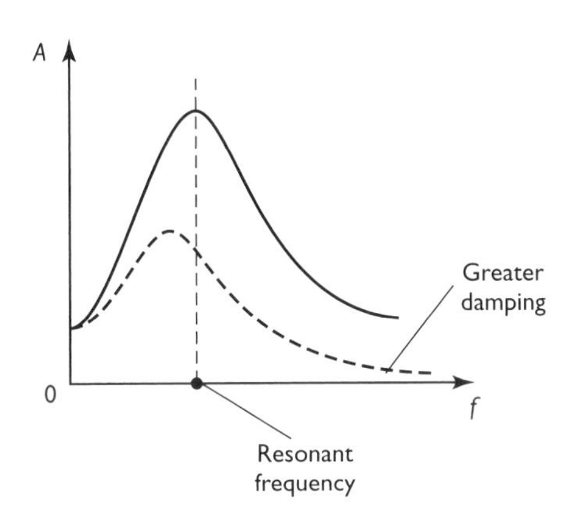

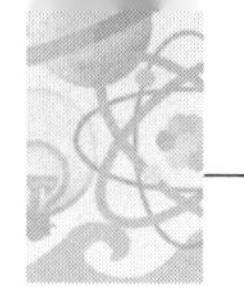

You should know the following points about a resonating system.

At resonance:

- the natural frequency of the forced oscillator (e.g. the bus seat) is equal to the forcing frequency (e.g. the engine)
- the forced oscillator has its maximum amplitude
- the forced oscillator absorbs maximum energy from the external source
- the degree of damping affects both the resonant frequency and the amplitude of the forced oscillator; greater degree of damping slightly reduces the resonant frequency (see the dotted graph above)

Resonance can be useful (e.g. microwave cooking, magnetic resonance imaging) or a nuisance (e.g. unnecessary vibration of car parts).

Solids, liquids and gases

Simple kinetic model

The three states of matter are solid, liquid and gas. All matter consists of particles, either atoms (e.g. metals) or complex molecules (e.g. DNA). Above a temperature of 0 K, the atoms or molecules in all matter are in some kind of motion.

- **Solid**: the molecules of a solid vibrate randomly about their equilibrium positions. The molecules are closely packed and they exert electrical forces on each other.
- **Liquid**: the molecules of a liquid have translational kinetic energy and they move randomly because of collisions with the other molecules. The mean separation between the molecules is slightly greater than in the solid state.
- **Gas**: the molecules have more pronounced random motion and the molecules have translational kinetic energy. The mean kinetic energy of the molecules increases with temperature (see page 53). The mean separation between the molecules is much greater than in the liquid state and depends on the size of the container. The molecules exert negligible electrical forces on each other, except when they collide.

The random motion of air molecules can be deduced by observing *Brownian motion* of smoke particles suspended in air. The smoke particles mimic the random motion of the microscopic air molecules.

Pressure

You should know the definition of pressure from Unit G481. A gas in a container exerts pressure on the container walls. This pressure arises from the numerous molecular collisions with the container walls. The kinetic model, together with Newton's laws of motion, can be used to explain the origin of pressure.

- A single molecule collides elastically with the container wall. For a molecule colliding normally (at 90°) with the wall, it rebounds with the same speed v. The magnitude of the change in momentum of the molecule is equal to $2mv$, where m is the mass of the molecule.

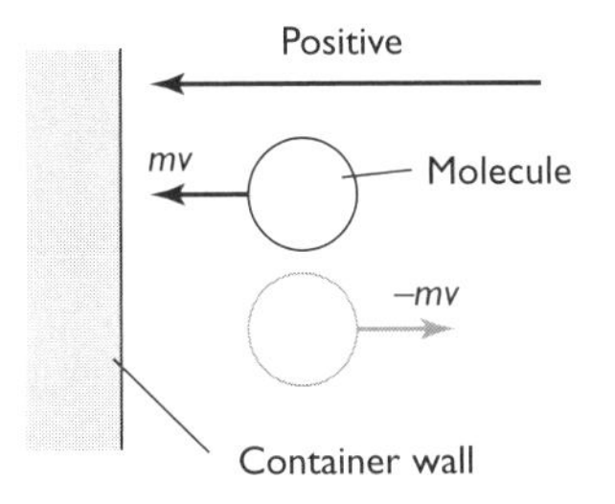

- A single molecule makes repeated collisions with the wall. The force on the molecule due to the wall is given by Newton's second law.
- The force on the molecule = rate of change of momentum = $\frac{\Delta p}{\Delta t} = \frac{2mv}{t}$ where t is the time between successive collisions with the wall.
- According to Newton's third law, the colliding molecule exerts an equal but opposite force on the wall.
- The total force acting on the wall is due to a large number of molecules colliding with the wall.

The pressure acting on the wall can be found using the following equation:

$$\text{pressure} = \frac{\text{total force}}{\text{cross-sectional area of wall}}$$

Worked example

Use the kinetic model to explain the change to the pressure exerted by gas in a container when:

(a) its temperature is increased

(b) its volume is decreased

Answer

(a) Increasing the temperature makes the molecules travel faster so their mean speed increases. When each molecule collides with the container wall, the change in momentum ($2mv$) is greater because the speed has increased. Since force is equal to the rate of change of momentum, the force each molecule exerts on the wall is therefore greater. Therefore, the pressure exerted by the gas increases.

(b) Decreasing the volume implies that molecules have less distance to travel between walls. This means that, on average, the time t taken between collisions of each molecule and the container wall decreases.

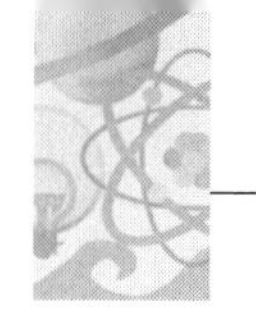

The change in momentum of the molecule is the same, but there is an increase in the 'rate of change of momentum'. The force exerted on the wall increases and hence the pressure also increases.

Internal energy

The molecules of a substance have kinetic energy (due to vibrations or translational motion) and potential energy (due to the electrical attraction between the molecules). As already mentioned, the molecules either move from one place to another randomly (in a liquid or gas) or vibrate randomly (in a solid). The internal energy of a substance is the total amount of energy of all the molecules of the substance.

The internal energy of a substance is the sum of the random distribution kinetic and potential energies of all the atoms or molecules.

The internal energy depends on the temperature of the substance. At 0 K (around −273°C) all molecules stop moving or vibrating. The internal energy of the substance is *minimum* and is entirely due to electrical potential energy between the molecules. The internal energy of a substance increases as its temperature is increased because the molecules gain kinetic and potential energy.

Melting, boiling and evaporation

The terms melting, boiling and evaporation have precise meaning in physics. Candidates often lose easy marks for poorly written definitions.

- **Melting**: describes when a substance changes from a solid state to a liquid state. A particular solid always melts at the same temperature. For example, pure ice melts at 0°C.
- **Boiling**: describes used when a substance changes from a liquid to a gaseous state. A particular liquid at a given pressure always boils at the same temperature. For example, pure water at standard pressure of 1.0×10^5 Pa boils at 100°C.
  During both melting and boiling:
  - the mean separation between the molecules increases
  - external energy is used to break molecular bonds
  - the electrical potential energy between the molecules increases
  - the mean kinetic energy of the molecules remains the same
  - the internal energy increases
- **Evaporation**: in this process, fast-moving molecules escape from the surface of a liquid. This leaves behind the slower-moving molecules — hence evaporation leads to cooling of a liquid. Unlike boiling, evaporation occurs at *all* temperatures of the liquid. The rate of evaporation of a liquid can be increased by both blowing over its surface and increasing its temperature.

Energy changes

The diagram below shows a typical graph of temperature against time for a solid substance (e.g. ice) heated at a constant rate.

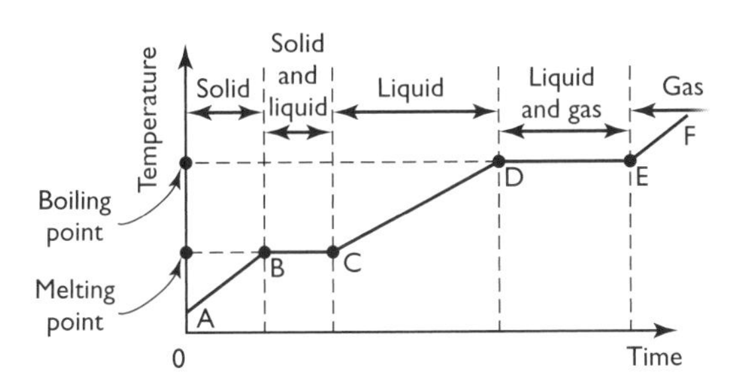

Section AB

The substance is in a *solid* state. As the temperature of the substance increases, the vibration energy of the molecules increases. There is little change in the separation between the molecules. The internal energy of the substance increases mainly because the molecules vibrate with greater amplitude and thus have more kinetic energy.

Section BC

The substance changes from *solid* to *liquid*. Energy is used to break molecular bonds and the mean separation between the molecules increases. The electrical potential between the molecules increases. The temperature remains constant: hence there is no increase in the kinetic energy of the molecules. (For ice, this takes place at 0° C.) The internal energy of the substance increases mainly because of the increase in electrical potential energy of the molecules. (At B the substance is entirely solid and at C it is entirely liquid.)

Section CD

The substance is in a *liquid* state. As the temperature of the substance increases, the mean translational kinetic energy of the molecules increases. There is little change in the separation between the molecules. The internal energy of the substance increases mainly because the molecules move faster and have more kinetic energy.

Section DE

The substance changes from *liquid* to *gas*. (For water, the gaseous state is known as 'water vapour'.) Energy is used to break bonds between molecules and the mean separation between the molecules increases considerably. The electrical potential between the molecules increases. The temperature remains constant, so there is no

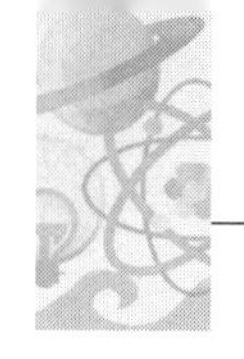

increase in the kinetic energy of the molecules. (For pure water, this takes place at 100°C at an external pressure of 1.0×10^5 Pa.) The internal energy of the substance increases mainly because of the increase in electrical potential energy of the molecules. (At D, the substance is entirely liquid and at C it is entirely gaseous.)

Section EF

The substance is a *gas*. Increasing the temperature increases the mean translational kinetic energy of the molecules. The electrical potential between the molecules is zero. The internal energy of the substance increases mainly because of the increased speed of molecules, resulting in increased kinetic energy.

Note: electrical potential energy between molecules can be quite confusing. The maximum electrical energy between molecules is zero when the molecules exert negligible electrical forces on each other because of the vast separation. When molecules are close together and exert forces on each other, the electrical potential energy is negative.

Temperature

Thermal equilibrium

When a hot object is in contact with a cooler object, there is a net heat transfer from the hot object to the cooler object. As a result, the cooler object gets warmer. Eventually, both objects will have the same temperature with no net flow of energy between the objects. The two objects in contact with each other are in **thermal equilibrium** when they have the same temperature. The diagram below shows the variation of temperature when a hot pizza at a temperature of 80°C is placed on a cooler plate at a temperature of 20°C.

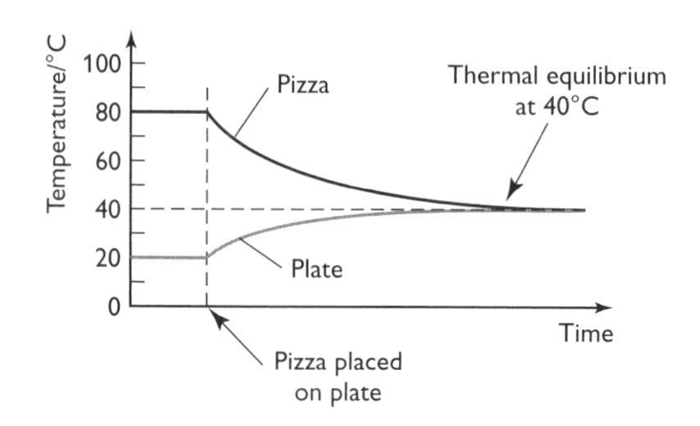

The pizza and the plate are in thermal equilibrium when they have the same temperature.

Temperature scales

You need to be familiar with the Celsius and the thermodynamic (Kelvin) temperature scales. For the Celsius scale, the unit is 'degrees Celsius' (°C) and for the thermodynamic scale the unit is kelvin (K). The letter θ is used for temperature on the Celsius scale and the letter T for temperature on the thermodynamic scale. The equations below show how to convert temperature between these two scales:

$$\theta\ (°C) = T\ (K) - 273.15$$

$$T\ (K) = \theta\ (°C) + 273.15$$

In most cases, you can use 273 instead of 273.15.

Note: A temperature of 0°C is equal to about 273 K. A change of 1°C is equal to a change of 1 K. At a temperature of 0 K, a substance has minimum internal energy.

Thermal properties of materials

Specific heat capacity

The specific heat capacity c of a substance is defined as follows:

The specific heat capacity of a substance is the energy required per unit mass of the substance to raise its temperature by 1 K (or 1°C).

In exams, it is easier to write the word equation for specific heat capacity:

$$\text{specific heat capacity} = \frac{\text{energy supplied}}{\text{mass} \times \text{temperature change}}$$

or

$$c = \frac{E}{m\Delta\theta}$$

where c is the specific heat capacity, E is the energy supplied to the substance, m is its mass and $\Delta\theta$ is the change in temperature. Specific heat capacity has the unit $J\ kg^{-1}\ °C^{-1}$ or $J\ kg^{-1}\ K^{-1}$.

Note: the temperature of a substance increases when external energy is supplied (as long as it does not change state). Its internal energy increases as the mean kinetic energy of each atom (or molecule) increases.

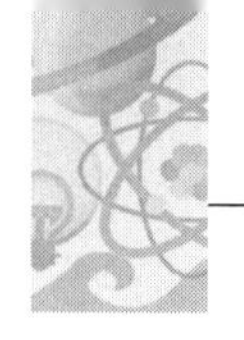

Worked example

A lagged beaker contains 80 g of water at a temperature of 20°C. The glass bulb of a 24 W filament lamp is immersed into the water. The temperature of the water doubles after a time of 5.0 minutes. The specific heat capacity of water is 4200 J kg^{-1} $°C^{-1}$. Assuming negligible heat loss to the surroundings, calculate the efficiency of the filament lamp as a heater.

Answer

The final temperature of water is 40°C.

energy supplied to heat water = $mc\Delta\theta$

$= 0.080 \times 4200 \times (40 - 20)$

$= 6.72 \times 10^3$ J

total energy supplied by lamp $= 24 \times (5.0 \times 60)$

$= 7.20 \times 10^3$ J

$$\text{efficiency} = \frac{6.72 \times 10^3}{7.20 \times 10^3} \approx 0.93$$

The filament lamp is 93% efficient as a heater; it provides 93% of its energy as heat and the rest (7%) as light.

Note: alternatively, you can determine the efficiency by using the equation:

$$\text{efficiency} = \frac{mc\Delta\theta}{Pt}.$$

Determining specific heat capacity

The diagram below shows the apparatus that can be used to determine the specific heat capacity of either a liquid or a solid.

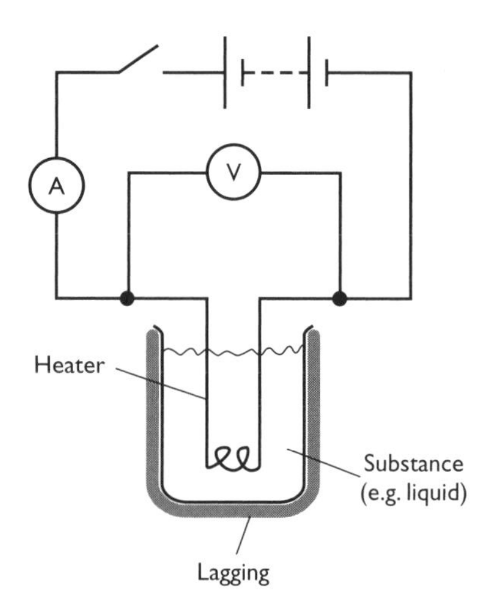

An electrical heater is used to heat the substance. Its temperature is measured using a thermometer. For a liquid, it is important to stir it before taking the temperature. The ammeter is connected in series and the voltmeter is connected in parallel with the heater. The substance is lagged to minimise heat losses to the surroundings.

Procedure

- Measure the mass m of the substance using a digital balance.
- Measure the initial temperature θ_i of the substance using the thermometer.
- Close the switch and start the stopwatch.
- Measure the current I in the heater and the p.d. V across it.
- Calculate the power P of the heater: $P = VI$.
- After a while, open the switch and stop the stopwatch.
- Measure the final temperature θ_f of the substance.
- Record the time t of heating from the stopwatch.
- Calculate the specific heat c as follows:

$$c = \frac{E}{m \times (\theta_f - \theta_i)} = \frac{VIt}{m \times (\theta_f - \theta_i)}$$

Latent heat of fusion and vaporisation

There is no temperature change when a substance changes state. The total kinetic energy of the molecules remains constant but their electrical potential energy increases as molecular bonds are broken. The term used for the energy supplied to change the state of a substance is 'latent heat'. The word latent means 'hidden'.

- **Latent heat of fusion** is the energy supplied to *melt* a solid substance to liquid.
- **Latent heat of vaporisation** is the energy supplied to *boil* a liquid substance to gas (or vapour).

To melt 1.0 kg of ice at 0°C to water at 0°C requires 0.33 MJ of energy. Do not forget that the same amount of energy is released when water changes to ice. To boil 1.0 kg of water at 100°C to water vapour at 100°C requires 2.26 MJ of energy. You can get severe burns from hot steam because of the high amount of heat released as it condenses into water. The diagram below summarises the energy required to change 1.0 kg of ice at 0°C into steam at 100°C.

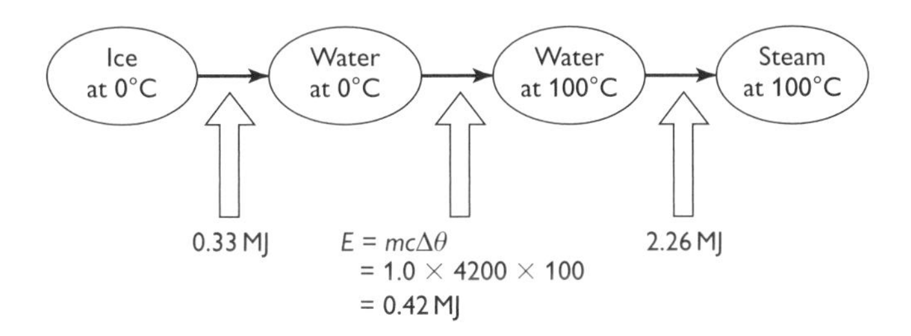

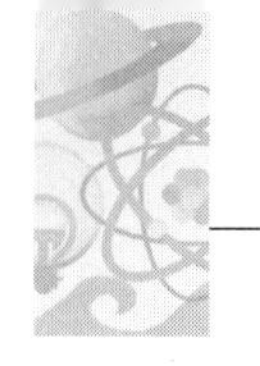

Ideal gases

The mole and the Avogadro constant

The mole is a mysterious concept to many physics candidates. Here are some key ideas about the mole.

- The amount of matter can be measured in either kilograms or moles.
- Measuring the amount of matter in moles is helpful because one mole of any substance has the same number of particles (atoms or molecules).
- One mole (often abbreviated to 'mol') of any substance has 6.02×10^{23} particles.
- The number 6.02×10^{23} is known as **Avogadro constant**, N_A.

One mole is defined as the amount of substance that has the same number of particles as there are atoms in 12 g of carbon-12 isotope.

For example:

- Carbon consists of atoms — 1 mole of carbon contains N_A atoms, 2 moles of carbon contain $2N_A$ atoms etc.
- Oxygen gas consists of molecules — 1 mole of oxygen gas contains N_A molecules, 2 moles of oxygen contain $2N_A$ molecules etc.

Worked example

Helium gas consists of atoms. The molar mass of helium is 4.0 g mol^{-1}. Determine the mass of a single helium atom.

Answer

1 mol has 6.02×10^{23} atoms of helium. The mass of 1 mol is 4.00×10^{-3} kg.

Therefore, mass of a helium atom = $\frac{4.00 \times 10^{-3}}{6.02 \times 10^{23}} = 6.64 \times 10^{-27}$ kg.

Boyle's law

The behaviour of a gas can be modelled using its volume V, the pressure p it exerts and its thermodynamic temperature T. Boyle's law is a statement of the behaviour of a fixed amount of gas at constant temperature:

The pressure exerted by a fixed amount of gas is inversely proportional to its volume, provided its temperature remains constant.

This can be written mathematically as:

$$p \propto \frac{1}{V}$$

or

$$pV = \text{constant}$$

For a gas at constant temperature, the product of pressure and volume remains constant. In simple terms this means that doubling the pressure will halve the volume. Similarly, decreasing the pressure by a factor of 10 will increase the volume of the gas by a factor of 10 etc. This leads to the following equation:

$$p_1V_1 = p_2V_2$$

where the subscript 1 represents the initial state of the gas and the subscript 2 represents the final state of the same gas.

Note: the equation above applies to ideal gases. Hydrogen, helium, oxygen are ideal gases at standard temperature and pressure (s.t.p.). However, as the temperature is lowered or the pressure becomes extremely high, some of these gases depart from their ideal behaviour because the electrical forces between the atoms (or molecules) are no longer negligible. Boyle's law cannot be used for real gases. In exams, you always assume that questions are about ideal gases.

The ideal gas equations

Experiments on ideal gases show that:

$$p \propto T \text{ for a gas at constant volume}$$

and

$$V \propto T \text{ for a gas at constant pressure}$$

Remember that the temperature T is in kelvin (K) and not °C. Combining the above relationships with Boyle's law gives the following relationship:

$$\frac{pV}{T} = \text{constant}$$

As before, this may be written as:

$$\frac{p_1V_1}{T_1} = \frac{p_2V_2}{T_2} \ (= \text{constant})$$

The constant in the relationship above depends on the number of moles (n) and the molar gas constant R, which has a value of $8.31\ \text{J K}^{-1}\ \text{mol}^{-1}$.

Therefore:

$$\frac{pV}{T} = nR$$

or

$$\boldsymbol{pV = nRT}$$

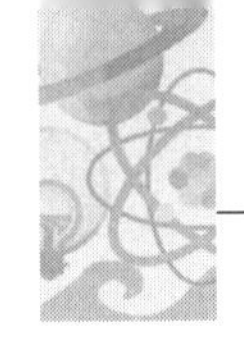

This equation is known as the **ideal gas equation** (or the **equation of state**). There is another version of the ideal gas equation that involves the number N of atoms (or molecules) of the gas and Boltzmann constant k, which has a value 1.38×10^{-23} J K^{-1}:

$$pV = NkT$$

The Boltzmann constant k is related to the molar gas constant R and Avogadro constant N_A by the equation:

$$k = \frac{R}{N_A}$$

In exams, it is vital that you can handle the equations in this section. The two most common mistakes are:

- confusing the number n of mols with the number N of atoms (or molecules)
- not converting the temperature into kelvin (K)

Worked example

Oxygen consists of O_2 molecules. The molar mass of oxygen is 32 g mol^{-1}. For 160 g of oxygen, calculate:

(a) the number of mols

(b) the number of molecules

(c) the volume occupied by the gas at a temperature of 20°C and pressure 1.01×10^5 Pa

Answer

(a) $n = \frac{160}{32} = 5.0$ mols

(b) The number of molecules of oxygen in 1 mol is 6.02×10^{23}, therefore

N = number of mols × Avogadro constant.

$N = 5.0 \times 6.02 \times 10^{23} = 3.01 \times 10^{24}$

(c) $pV = nRT$, where the temperature T is in kelvin.

$$V = \frac{nRT}{p} = \frac{5.0 \times 8.31 \times (273 + 20)}{1.01 \times 10^5}$$

$V = 0.1205\ \text{m}^3 \approx 0.12\ \text{m}^3$

n Note: You could also use $pV = NkT$ to answer part (c).

Translational kinetic energy

n Note: this section refers to atoms, but these could of course be molecules.

Increasing the temperature of the gas makes the atoms travel faster and their mean translational kinetic energy increases. The mean kinetic energy of the atoms is related to the thermodynamic temperature T of the gas:

mean translational kinetic energy of atoms ∝ T

The internal energy of a gas is almost entirely equal to the total kinetic energy of the atoms, so internal energy $\propto T$.

For individual atoms, the mean translational kinetic energy E is given by the equation:

$$\boldsymbol{E = \frac{3}{2}kT}$$

where k is the Boltzmann constant. The mean kinetic energy E is equal to $\frac{1}{2}mv^2$, where m is the mass of the atom and v is the mean speed of the atoms. At a given temperature, the gas atoms have a random motion and travel at different speeds. However, the mean kinetic energy of the atoms remains constant at a given temperature. The graph below shows the variation of the kinetic energy of gas atoms at a given temperature.

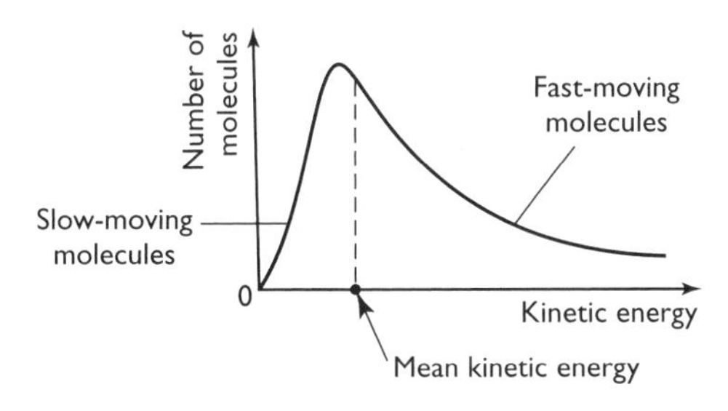

Worked example

The outer surface of a star has hydrogen and helium *atoms*. The surface temperature of the star is about 8000 K. The molar mass of hydrogen is $1.0\ \text{g mol}^{-1}$ and the molar mass of helium is $4.0\ \text{g mol}^{-1}$.

(a) Explain whether the hydrogen and helium atoms have the same or different mean kinetic energy.

(b) Calculate the mean speed of the hydrogen atoms.

(c) Determine the ratio $\dfrac{\text{mean speed of hydrogen atoms}}{\text{mean speed of helium atoms}}$.

Answers

(a) The mean translational kinetic energy E depends on the temperature T, that is:

$$E = \frac{3}{2}kT \propto T$$

Since the temperature is the same for hydrogen and helium, their mean kinetic energy is the same.

(b) $E = \frac{3}{2}kT$

$$\frac{1}{2}mv^2 = \frac{3}{2}kT$$

$$v = \sqrt{\frac{3kT}{m}}$$

The mass m of a single hydrogen atom is:

$$m = \frac{1.0\times10^{-3}}{6.02\times10^{23}} = 1.661\times10^{-27}\ \text{kg}$$

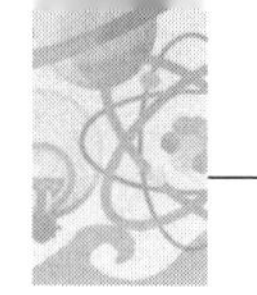

Therefore:

$$v = \sqrt{\frac{3 \times 1.38 \times 10^{-23} \times 8000}{1.661 \times 10^{-27}}} \approx 1.4 \times 10^{4}\ \text{m s}^{-1} \qquad (14\ \text{km s}^{-1})$$

(c) The mean speed v is inversely proportional to $\sqrt{m}$. The mass of the helium atom is 4 times greater than the mass of the hydrogen atom. Since $\sqrt{4} = 2$, the hydrogen atoms will have twice the mean speed of helium atoms. Therefore:

$$\text{ratio} = \frac{\text{mean speed of hydrogen atoms}}{\text{mean speed of helium atoms}} = 2$$

&

This section contains questions similar in style to those you can expect on the exam paper for Unit G484. The responses of two candidates are given. The answers from Candidate A are similar to those expected from a student who has reached grade-A* standard.

The answers given by Candidate B are incomplete or inappropriate answers and would earn a grade C.

At least one question is given for each section of this unit. The Question and Answer section can be used in several ways. The best way to engage your brain is to write something on a piece of paper. There is no value in just reading through the questions and answers given by the two candidates. Here are some possible strategies:

- Do the question yourself and mark it using the comments from the examiner. You can then compare the responses of two candidates and learn from your mistakes.
- Mark the answers given by the two candidates and check whether you agree with the mark awarded by the examiner.

Examiner comments

Every mark scored by a candidate is shown by a tick at the relevant place. The total number of ticks (✓) and crosses (✗) should add up to the total mark for the question. All candidates' responses are followed by examiner comments. These are denoted by the icon e. These comments focus on the errors made by candidates and sometimes offer alternative ways of securing the marks available. The comments at the end of each question provide valuable tips for answering examination questions.

Question 1

Newtonian laws of motion

(a) Define the term *linear momentum* and explain why it is a vector quantity. (2 marks)

(b) Write a word equation for Newton's second law. (1 mark)

(c) A ball of mass 350 g falls vertically and hits the ground at a speed of 5.2 m s^{-1}. It rebounds with a speed of 4.0 m s^{-1}. The ball is in contact with the ground for 80 ms.

(i) Calculate the magnitude of the change in momentum of the ball. (2 marks)

(ii) Calculate the magnitude of the average force exerted by the ground on the ball. (2 marks)

(iii) State and explain the value of the average force exerted by the ball on the ground. (2 marks)

(d) A rocket engine ejects 600 kg of exhaust gases in each second at a speed of 360 m s^{-1} relative to the rocket. Calculate the force exerted by the exhaust gases on the rocket. (3 marks)

Total: 12 marks

Candidates' answers to Question 1

Candidate A

(a) linear momentum = mass × velocity of object ✓
It has the unit kg m s^{-1}.
Momentum is a vector because it is a product of a scalar (mass) and a vector (velocity). ✓

Candidate B

(a) $p = mv$
where p is the momentum, m is the mass and v is velocity (and not speed). ✓
Momentum is a vector because it has direction and size. ✗

Candidate A has made an excellent start with a word equation for momentum and shown a clear understanding of why momentum is a vector quantity. Candidate B has done well with the definition but has not answered the question fully.

Candidate A

(b) Newton's second law: the net force acting on an object is directly proportional to the rate of change of momentum. The force is in the same direction as the change in momentum. ✓

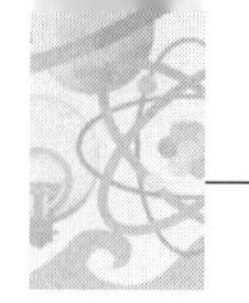

Candidate B

(b) The total force acting on an object is equal to the rate of change of momentum.✓

Each candidate states the second law correctly. It is worth noting that you can use either 'equal to' or 'directly proportional to' in stating this law.

Candidate A

(c) (i) $p = mv$

initial momentum = +(0.350 × 5.2) = +1.82 kg m s^{-1} ✓

final momentum = −(0.350 × 4.0) = −1.40 kg m s^{-1}

Δp = −1.40 − 1.82 = −3.22 kg m s^{-1}

Therefore, the change in momentum has a magnitude of 3.22 kg m s^{-1}. ✓

Candidate B

(c) (i) Δp = (0.350 × −4.0) − (0.350 × 5.2) = −3.22 ✓

The change in momentum is about 3.2 kg m s^{-1}. ✓

Both candidates give good answers that show they understand the vector nature of momentum.

Candidate A

(c) (ii) force $= \frac{\Delta p}{\Delta t} = \frac{3.22}{0.080} = 40.25$ N ✓✓

force ≈ 40 N

Candidate B

(c) (ii) $F = ma = 0.350 \times \left(\frac{-4.0 - 5.2}{0.080}\right)$ ✓

force = 40.25 N ✓

Candidate A uses the value for the change in momentum to determine the force. Candidate B uses $F = ma$ and the values of the velocities. Both strategies are equally acceptable. Candidate B should have quoted the value of the force to two significant figures, but examiners tend to be lenient with more significant figures.

Candidate A

(c) (iii) The force on the ground is also equal to 40 N but in the opposite direction. ✓

This is because of Newton's third law of motion. ✓

Candidate B

(c) (iii) The force is also 40.25 N. ✓

The question is worth 2 marks, with Candidate B stating the correct value of the force but not providing any explanation.

Candidate A

(d) force $= \frac{\Delta(mv)}{\Delta t} = \frac{\Delta m}{\Delta t} \times v$ ✓

force = 600 kg s^{-1} × 360 m s^{-1} = 2.16 × 10^{5} N ✓✓

Candidate B

(d) momentum = 600 × 360 = 216000 ✓✗

Candidate A gives a perfect answer. Candidate B calculates the 'change of momentum per second' but does not realise that this is the same as the force acting on the rocket.

The presentation of Candidate A's answer is immaculate and an examiner would have no difficulty in following the answers and awarding full marks. Candidate B has picked up 8 marks and lost some easy marks for incomplete answers. The answers from this candidate were either too brief or incomplete. Just gaining one extra mark would have elevated Candidate B to a grade B.

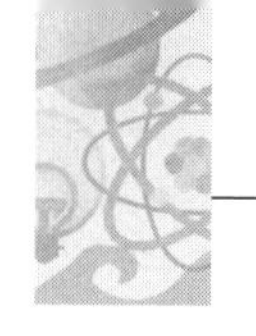

Question 2

Collisions

(a) **Explain what is meant by an *inelastic* collision.** (2 marks)

(b) **State the principle of conservation of momentum.** (1 mark)

(c) **A car accelerates from rest. It gains both velocity and momentum. Discuss whether or not the principle of conservation of momentum is being broken.** (2 marks)

(d) **The diagram below shows a block of wood of mass 450 g resting on the edge of a table.**

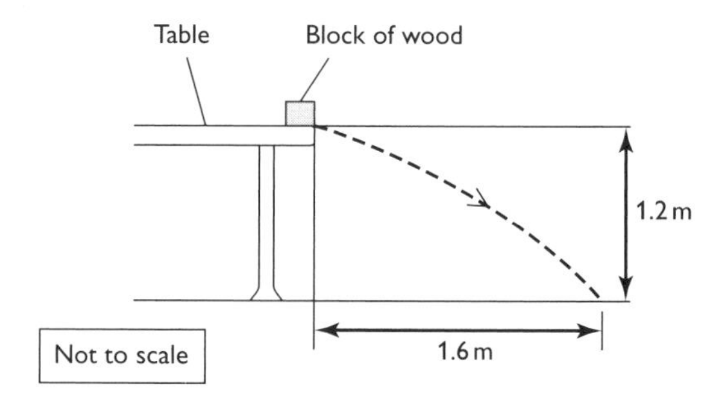

A bullet of mass 5.0 g is fired horizontally into the wood. The bullet remains in the block. The block, with the embedded bullet, falls off the edge of the table and travels a horizontal distance of 1.6 m before hitting the ground. The height of the table is 1.2 m.

(i) **Show that the final speed of the block, with the embedded bullet, is about 3.2 m s^{-1}. Assume air resistance has negligible effect on the motion of the block.** (3 marks)

(ii) **Hence, calculate the initial speed of the bullet.** (3 marks)

Total: 11 marks

Candidates' answers to Question 2

Candidate A

(a) In such a collision, both momentum and *total* energy are conserved. ✓✓

Candidate B

(a) Momentum is conserved in this collision. ✓
Kinetic energy is also conserved. ✗

e Candidate A gives a perfect answer. Candidate B secures 1 mark for realising that momentum is conserved. Kinetic energy is only conserved in an elastic collision, so he/she earns no mark for this wrong physics.

Candidate A

(b) In a closed system (no external forces), the momentum before and after a collision remains the same. ✓

Candidate B

(b) In a collision, we also have:
total initial momentum = total final momentum. ✓

Both candidates correctly recall this fundamental part of physics.

Candidate A

(c) When the car is a rest, the total momentum of the Earth and the car is zero. ✓
If the car has a forward momentum $+p$, the Earth has an equal momentum $-p$ in the opposite direction — making a total momentum of zero.
Hence, the Earth moves in the opposite direction to the car, with the same momentum. ✓

Candidate B

(c) The Earth must move in the opposite direction with the same momentum as the car. ✓

Candidate A provides a logical set of statements and the answer is detailed. Candidate B's response is too brief and does not mention the role played by the principle of conservation of momentum.

Candidate A

(d) (i) This is a question on projectiles. The horizontal velocity is constant and the vertical velocity changes according to the equations of motion.

Vertically: $s = \frac{1}{2}at^2$ ✓ $t = \sqrt{\left(\frac{2s}{a}\right)} = \sqrt{\left(\frac{2 \times 1.2}{9.81}\right)} = 4.94625\text{ s}$ ✓

Horizontally: speed $= \frac{1.6}{4.9462} = 3.23\text{ m s}^{-1}$ ✓

Candidate B

(d) (i) I can use '$s = ut + \frac{1}{2}at^2$' for vertical motion where s is 1.2 m.

To find the speed I can use '$v = \frac{1.6}{t}$' for the horizontal motion. ✓

Both candidates realise that this is a synoptic question requiring an understanding of projectiles. Candidate B does not substitute any values into the equations, but the examiner would award 1 mark for some sensible physics. Candidate A presents a flawless answer.

Candidate A

(d) (ii) initial momentum = final momentum

$5.0v = (450 + 5.0) \times 3.23$ ✓

$v = \frac{455 \times 3.23}{5.0} \approx 290\text{ m s}^{-1}$ ✓✓

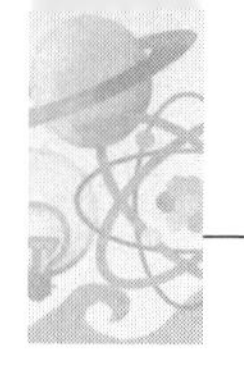

Candidate B

(d) (ii) $0.005 \times v = 0.455 \times 3.2$ ✓

$v = 291.2$ m/s ✓✓

Candidate A uses the calculated value 3.23 m s^{-1} to solve the problem. Working consistently in grams is also fine. Candidate B opts to work in SI units and uses the answer given in (d)(i) to calculate the speed of the bullet.

Candidate A scores full marks and shows an excellent understanding of physics — the answers are all detailed. Candidate B scores 7 marks and struggles with the synoptic question on projectiles in (d)(i). This candidate could easily improve his/her performance by providing detailed answers and learning basic definitions.

Question 3

Circular motion

(a) **The diagram below shows a sketch drawn by a student showing the direction of the centripetal force *F* acting on an object moving in a circle.**

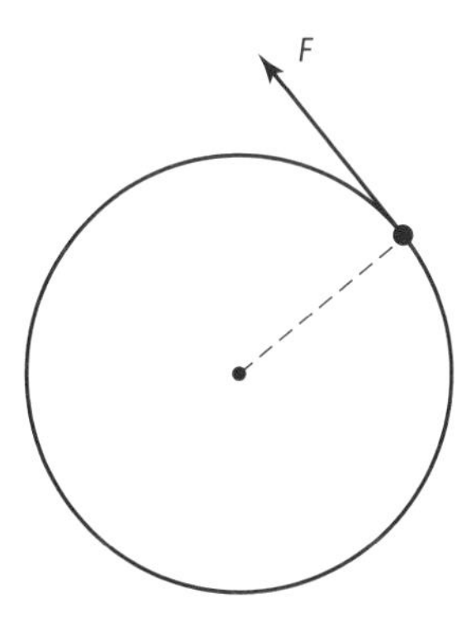

Explain what is wrong with this sketch. (1 mark)

(b) **Derive an equation for the constant speed *v* of object moving in a circle of radius r and having a period *T*.** (2 marks)

(c) **A 50 g mass is tied to length of string. It is whirled in a vertical circle of radius 0.40 m at a constant speed of 8.0 m s^{-1} (see diagram below).**

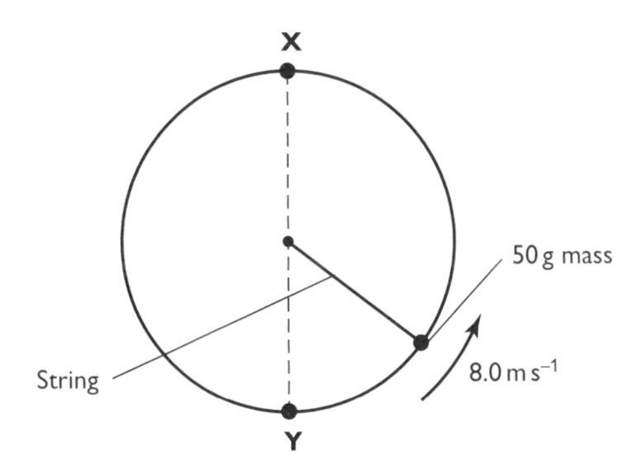

(i) **Calculate the weight of the object.** (1 mark)

(ii) **Compare the magnitudes of the tension at X and Y as the mass is whirled at a constant speed in a vertical circle.** (2 marks)

(iii) **Calculate the tension in the string when the mass is at the top of the circle.** (3 marks)

Total: 9 marks

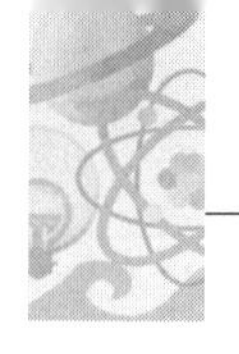

Candidates' answers to Question 3

Candidate A

(a) The direction of the centripetal force should always point towards the centre of the circle. ✓

Candidate B

(a) The arrow for F should be pointing towards the centre of the circle (and not a tangent). ✓

This is good start from both candidates. The centripetal force is always directed towards the centre. Perhaps the student's sketch in the question was for the direction of the velocity and not the centripetal force.

Candidate A

(b) speed $= \dfrac{\text{distance}}{\text{time}}$ ✓

The distance is equal to the circumference of the circle ($2\pi r$) and the time is the period (T). Therefore $v = \dfrac{2\pi r}{T}$. ✓

Candidate B

(b) v = circumference/period $= \dfrac{2\pi r}{T}$ ✓✓

Candidate A gives a clear proof for the speed v. Although Candidate B's answer is brief, the physics is clear.

Candidate A

(c) (i) weight $= mg = 0.050 \times 9.81 = 0.491\ \text{N} \approx 0.49\ \text{N}$ ✓

Candidate B

(c) (i) $W = 0.49\ \text{N}$ ✓

Candidate B does not shown his/her working, but the answer is correct for this 1-mark question. Candidate B also uses $g = 9.8\ \text{m s}^{-2}$, which is acceptable in exams.

Candidate A

(c) (ii) The centripetal force must always be towards the centre of the circle.

At X, the weight W and the tension T together provide the (constant) centripetal force.

Hence, the tension will be less than the weight at X. ✓

At Y, the tension must be greater than the weight. ✓

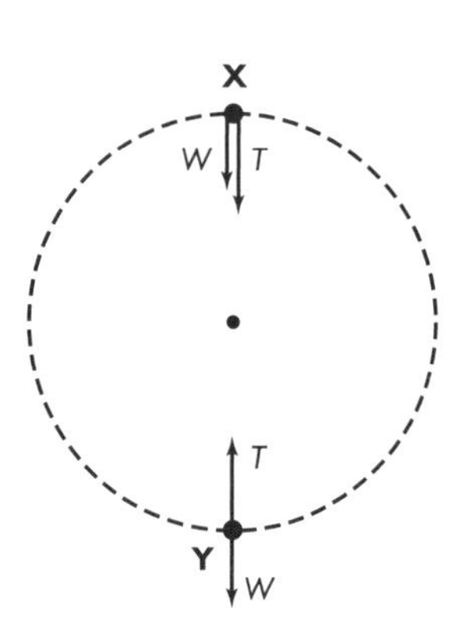

Candidate B

(c) (ii) The centripetal force is constant; hence, the weight is always equal to the tension. ✗✗

This is a difficult question, but Candidate A gives a good response. Candidate B fails to give a satisfactory answer.

Candidate A

(c) (iii) centripetal force $= \frac{mv^2}{r} = \frac{0.050 \times 8.0^2}{0.40} = 8.0$ N ✓✓

centripetal force = weight + tension = 8.0 N

Therefore: $T = 8.0 - 0.491 \approx 7.5$ N ✓

Candidate B

(c) (iii) $F = \frac{0.050 \times 8.0^2}{0.40} = 8.0$ newtons ✓✓

The tension must be the same as 8.0 N. ✗

Candidate B does well to secure 2 marks, especially earning no marks from (c)(ii). The tension in the string would be 8.0 N if the motion were in a horizontal circle. In a vertical circle, the tension changes in order to maintain a constant centripetal force. Candidate A's answer is perfect. Note that $mg + T = \frac{mv^2}{r}$ at the top of the circle.

Overall Candidate A secures full marks and demonstrates that grade-A* candidates do not have any gaps in their knowledge of physics. Candidate B scores 6 marks and does not appreciate that the tension in the string would change. Perhaps drawing a sketch in (c)(ii), like Candidate A, might have helped Candidate B to visualise the role of the tension and weight.

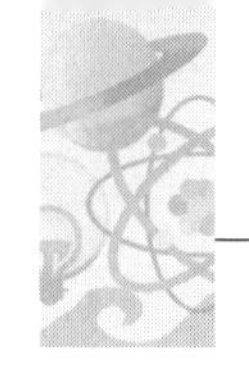

Question 4

Gravitational fields

(a) State Newton's law of gravitation. (2 marks)

(b) Define gravitational field strength at a point in space. (1 mark)

(c) The planet Mars has mass 6.4×10^{23} kg and a surface gravitational field strength of 3.8 N kg^{-1}.

(i) Calculate the diameter of Mars. (3 marks)

(ii) Determine the surface gravitational field strength on the surface of Jupiter, given its diameter is 21 times that of Mars and it is 2960 times more massive than Mars. (3 marks)

Total: 9 marks

Candidates' answers to Question 4

Candidate A

(a) The attractive force between two point masses is directly proportional to the product of their masses and inversely proportional to the square of their separation. ✓✓

Candidate B

(a) The attractive force between two particles is given by:

$$\text{force} \propto -\frac{\text{mass}_1 \times \text{mass}_2}{\text{separation}^2}$$ ✓✓

The answers may look different but both candidates communicate the physics effectively. This is a good start from both candidates. Candidate B would have secured 2 marks without the minus sign in the relationship, because he/she has already stated the force is attractive.

Candidate A

(b) Gravitational field strength g is the force acting on an object per unit mass. ✓

Candidate B

(b) gravitational field strength $= \frac{\text{force}}{\text{mass}}$ ✓

Too many candidates lose this easy mark in exams with a statement such as 'This is the force on a unit mass'. Here both candidates state clearly that gravitational field strength is the force on an object divided by its mass.

Candidate A

(c) (i) $g = \frac{GM}{r^2}$ ✓

radius = $\sqrt{\frac{GM}{g}}$

diameter = $2 \times \sqrt{\frac{GM}{g}} = 2 \times \sqrt{\frac{6.67 \times 10^{-11} \times 6.4 \times 10^{23}}{3.8}}$ ✓

diameter = 6.7×10^{6} m ✓

Candidate B

(c) (i) $g = \frac{GM}{r^2}$ ✓

$3.8 = \frac{6.67 \times 10^{-11} \times 6.4 \times 10^{23}}{r^2}$ ✓

$r = 1.123 \times 10^{13}$ m ✗

Candidate A provides a perfect answer that sets out the physics logically. The examiner would be generous in awarding 2 marks to Candidate B, who forgets to square-root his answer and double it to find the diameter — a careless error.

Candidate A

(c) (ii) $g \propto \frac{M}{r^2}$

g is proportional to mass, hence 2,960 times larger than 3.8 N kg^{-1}.

g is inversely proportional to square of distance, hence 21^2 times smaller than 3.8 N kg^{-1}.

$g = 3.8 \times \frac{2960}{21^2} = 25.5$ N kg^{-1} ✓✓✓

Candidate B

(c) (ii) mass = $2960 \times 6.4 \times 10^{23}$

radius = $21 \times 1.123 \times 10^{13}$

$g = \frac{GM}{r^2} = \frac{G \times (2960 \times 6.4 \times 10^{23})}{(21 \times 1.123 \times 10^{13})^2}$ ✓

g = ? ✗✗

Candidate A demonstrates again a good understanding and earns full marks. Candidate B could also have picked up 3 marks but fails to calculate the final value. Even though the value for the radius is wrong, examiners can award marks for the correct physics through the 'error carried forward' rule.

Candidate A scores full marks and demonstrates a good understanding of physics. Candidate B loses some marks unnecessarily and scores only 6 marks. He/she could have easily scored a few more marks by writing down all the stages of the calculations in (c). For example, the answer given in (c)(i) is r^2 and not the radius or the diameter. Candidate B could easily improve his/her performance by one grade by reading the questions carefully and not rushing through analytical work.

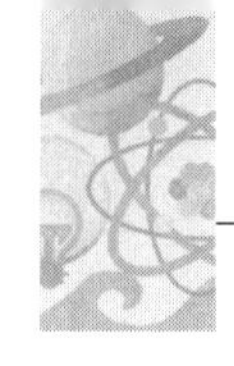

Question 5

Simple harmonic motion

A mass of 0.800 kg oscillates with simple harmonic motion on the end of a spring. The variation of its acceleration *a* with displacement *x* is shown in the graph below.

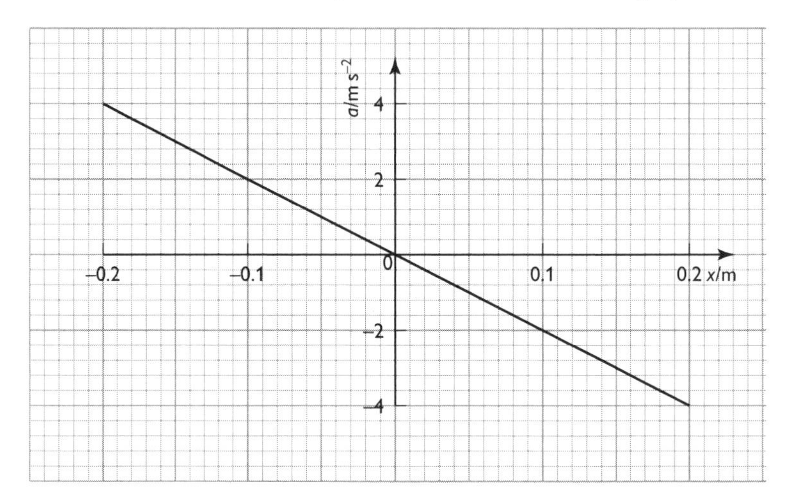

(a) Define *simple harmonic motion.* (2 marks)

(b) Explain how the graph above shows that the motion is simple harmonic. (2 marks)

(c) Use the graph above to determine:

(i) the amplitude of the mass (1 mark)

(ii) the frequency *f* of the oscillations (3 marks)

(d) The spring with the 0.800 kg mass is hung vertically from a mechanical oscillator. The frequency of the oscillator is slowly increased from zero. With the aid of a labelled sketch, describe the motion of the mass. (5 marks)

Total: 13 marks

Candidates' answers to Question 5

Candidate A

(a) The acceleration of the mass is always directed towards the equilibrium position and is proportional to the displacement. ✓✓ That is, $a \propto -x$.

Candidate B

(a) The acceleration is proportional to negative displacement. ✓

Candidate A has learnt the definition for simple harmonic motion. Candidate B gains 1 mark for stating the relationship between the acceleration and the displacement.

However, he/she does not explain what is meant by 'negative displacement'. At A2, examiners expect definitions to be word-perfect.

Candidate A

(b) The graph is a straight line through the origin. Hence, acceleration is proportional to the displacement. ✓
The gradient is negative — this shows that acceleration is towards some fixed point. ✓

Candidate B

(b) The graph is a straight line through the origin, therefore $a \propto x$. ✓

The answer from Candidate A is perfect. Candidate B's response is too brief and he/she should have realised that 2 marks warranted another statement. This candidate does not understand that the negative slope of the line means that the acceleration is always opposite to the displacement and hence is directed towards the equilibrium position.

Candidate A

(c) (i) The amplitude is 20 cm. ✓

Candidate B

(c) (i) $A = 0.2$ metres ✓

Both candidates realise that the maximum displacement from the graph is equal to 0.20 m.

Candidate A

(c) (ii) $a = -(2\pi f)^2 x$

The gradient must therefore be $(2\pi f)^2$. ✓

$$\text{gradient} = \frac{4.0}{0.2} = 20$$

$$2\pi f = \sqrt{20} \; ✓$$

$$f = \frac{\sqrt{20}}{6.284} = 0.71 \text{ Hz} \; ✓$$

Candidate B

(c) (ii) $a = -(2\pi f)^2 x$

$-4.0 = -(2\pi f)^2 \times 0.2$ ✓

$f = 0.71$ hertz ✓✓

Candidate A's answer is well laid out and all correct. Candidate B uses one point from the graph ($x = 0.20$ m, $a = -4.0$ m s^{-2}) to determine the frequency. In this case, there is nothing wrong with such a strategy — since the graph is a straight line through the origin. Unlike Candidate A, Candidate B has been foolish by not showing all the working for the final answer, but on this occasion, the answer was correct.

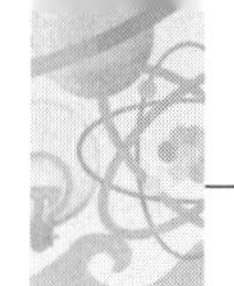

Candidate A

(d) I have drawn a graph of amplitude A of the mass against the forcing frequency f.

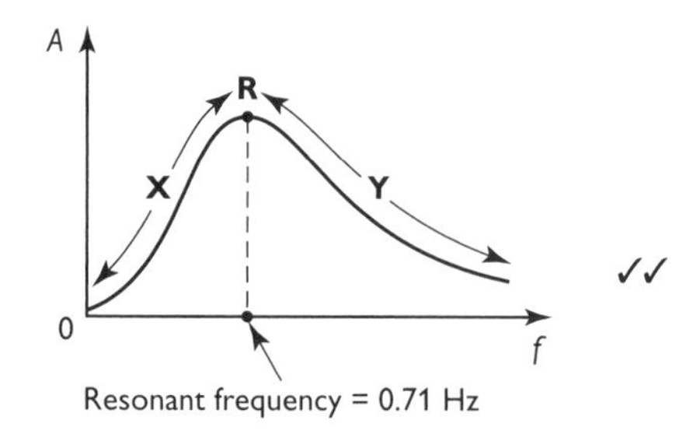

As frequency is increased from zero, the amplitude starts to increase (part X). ✓
At 0.71 Hz, the amplitude is a maximum and resonance takes place (R). ✓
As frequency is increased beyond 0.71 Hz, the amplitude decreases as shown on the graph part Y). ✓

Candidate B

(d) Here is my graph.

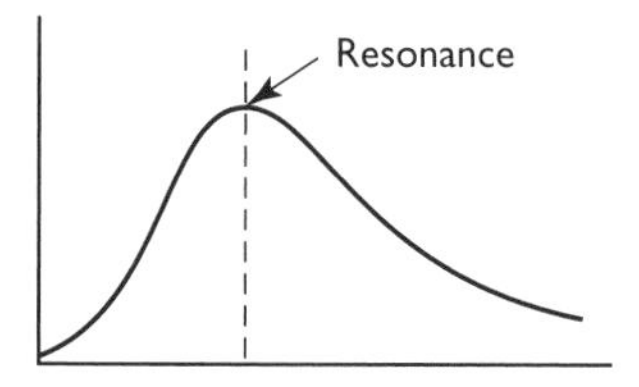

The mass will show resonance at 0.71 Hz and have maximum amplitude. ✓
The frequency also changes the amplitude.

e A labelled sketch must show what is being plotted and show values of any key features resonance in this case. Candidate A has shown how this is done. The description is also perfect. Candidate B has lost all marks for the sketch because the axes are not labelled. His/her description is also too brief.

e **Candidate A gives a model answer and scores full marks. All the necessary detail is given and he/she clearly understands the work thoroughly. The answer for (d) is particularly praiseworthy because it requires extended writing and sketching a graph. Candidate B loses marks because of poor recall of definitions and careless work. The description in (d) is too brief and the sketch graph would have taken valuable time to draw but scores no marks because the axes are not labelled. Sloppy work costs Candidate B at least one grade.**

Question 6

States of matter and temperature

(a) Explain what is meant by two objects being in *thermal equilibrium*. (1 mark)

(b) Define *internal energy* of a substance. (2 marks)

(c) When some substances are heated, they change state from a solid directly to a gas.

(i) Describe the motion of the molecules in the solid and gaseous states. (2 marks)

(ii) State and explain how the internal energy of the substance changes as it transforms from a solid to a gas. (2 marks)

(d) Use the kinetic theory model to explain how gas molecules in a container exert pressure. (4 marks)

Total: 11 marks

Candidates' answers to Question 6

Candidate A

(a) The two objects have the same temperature.
There is no net flow of heat (thermal energy) between the objects. ✓

Candidate B

(a) The objects must be at the same temperature. The amount of energy gained by one object is equal to the energy transferred to the other object. ✓

There is no mark here for the temperature being the same. Both candidates do well to state that there is effectively no net transfer of energy between the objects.

Candidate A

(b) This is the sum of kinetic energy and electrical potential energy of all the molecules. ✓ The energies of the molecules are random. ✓

Candidate B

(b) This is the total KE and PE of all the particles (atoms or molecules). ✓

This is a 2-mark question and Candidate B's answer is brief and does not mention that energy of the particles is randomly distributed.

Candidate A

(c) (i) The molecules in a solid vibrate randomly about fixed points. ✓
The molecules of a gas have kinetic energy. They move randomly through space and have a range of velocities. The molecules have translational kinetic energy. ✓

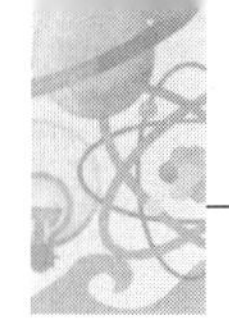

Candidate B

(c) (i) The particles (molecules or atoms) in a solid vibrate and have vibrational kinetic energy. ✓ The gas particles do not exert a force on each other. The particles move randomly and have translational kinetic energy. ✓

Both candidates clearly state the main characteristics of the particles in a solid and in a gas.

Candidate A

(c) (ii) In a solid, the internal energy is due to the vibrational energy of the molecules and the electrical energy of the molecules. As the particles change state, the internal energy of the substance increases. ✓ The potential energy of the molecules increases — molecular bonds are broken. ✓

Candidate B

(c) (ii) It increases as energy is supplied to the solid. ✓

Such exam answers can be riddled with misconceptions. Candidate A has an excellent understanding of internal energy. Candidate B has most likely guessed but does secure a mark for the correct answer. However, there is no attempt to explain the answer.

Candidate A

(d) The molecules collide with the container wall. ✓ For each collision, there is a change in momentum equal to twice the initial momentum of the molecule. ✓ There is a force on the molecule as there is rate of change of momentum. ✓ According to Newton's third law, each molecule exerts an equal but opposite force on the wall. ✓ The pressure is equal to the total force due to all the colliding molecules divided by the area of the wall.

Candidate B

(d) pressure = force/area ✓

The molecules or molecules colliding with the wall provide the force. ✓ I can use Newton's laws to explain this.

Examiners often have more marking points than the actual marks, so it is always safer to write more rather than less. Candidate A could have scored for the last statement, but had already scored the maximum 4 marks. Candidate B makes the mistake of being too brief.

Candidate A scores maximum marks. His/her answer typifies the work of grade-A* candidates. He/she clearly understands the questions and gives detailed answers. Candidate B scores 7 marks and would earn a grade C. This candidate needs to scrutinise each question carefully before writing the answers. There are too many gaps in his/her knowledge. The number of marks for a question often indicates how much detail is required in the answer.

Question 7

Thermal properties of materials

(a) Define *specific heat capacity* of a substance. (1 mark)

(b) Describe an electrical experiment to determine the specific heat capacity of a metal block. (5 marks)

(c) An electric kettle is rated as 2.5 kW. The kettle is filled with 0.90 kg of water at a temperature of 12°C. Calculate the time taken for the kettle to boil. Assume all the energy supplied by the heater is used to warm the water. The specific heat capacity of water is 4200 J kg^{-1} K^{-1}. (3 marks)

Total: 9 marks

Candidates' answers to Question 7

Candidate A

(a) The specific heat capacity c is given by the equation $c = \frac{E}{m\Delta\theta}$, where E is the energy supplied to the substance, m is its mass and $\Delta\theta$ is the change in temperature. ✓

Candidate B

(a) A substance has specific heat capacity. It tells us how much heat is needed to change its temperature. ✗

There is really no excuse for not learning definitions at A2. Candidate B has given an extremely vague statement and has made a poor start on this easy opening question.

Candidate A

(b) I would provide the metal with energy from a heater. An ammeter is connected in series with the heater and the voltmeter is placed across it. ✓ The power of the heater is found from the ammeter and voltmeter readings. The power P of the heater is 'voltage × current'. ✓ The metal block is heated for a time t (found using a stopwatch). The mass of the metal block is found using a digital balance. ✓ A thermometer can be used to record the initial and final temperatures. The change in temperature $\Delta\theta$ is found by subtracting these readings. ✓ The specific heat capacity is found using the equation $c = \frac{Pt}{m\Delta\theta}$. ✓

To make the experiment accurate, I would lag the metal block and repeat the experiment twice.

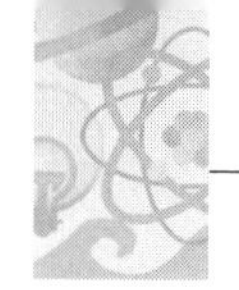

Candidate B

(b) A heater of known power P is used (e.g. 25 W).
I would heat the block for a certain time t and find the energy E supplied to the metal block using the equation $E = Pt$. ✓ The specific heat capacity is found using $c = \frac{E}{m\Delta\theta}$. ✓ The mass of the metal block is found using a balance or scales. ✓ The temperature can be recorded using a data-logger or a thermometer.

Examiners always have more marking points than the marks for such a descriptive question. Candidate A is methodical in describing the experiment. Notice how much detail the answer contains. Candidate B's answer is brief. There is no indication of how $\Delta\theta$ is determined or how the power of the heater may be determined using meters.

Candidate A

(c) $Pt = mc\Delta\theta$
Water boils at a temperature of 100°C. Therefore:
$2500 \times t = 0.90 \times 4200 \times (100 - 12)$ ✓
$t = \frac{0.90 \times 4200 \times 78}{2500}$ ✗
$t = 118$ s ✓ *(error carried forward)*

Candidate B

(c) energy = $0.90 \times 4200 \times 88 = 332640$ J ✓
time = $\frac{332640}{2500} = 133$ s ✓✓

Candidate A loses 1 mark for the wrong value for the change in temperature (it should be 88°C and not 78°C). The examiner deducts 1 mark for this error and applies the 'error carried forward' (ECF) rule for the answer for time. This shows that even a grade-A candidate can make a slip. On this occasion, Candidate B provides a perfect answer.

Candidate A scores 8 marks and is still on target for a grade A. The answers are detailed, especially for the description of the experiment in (b). Candidate B scores 6 marks and could easily have picked up extra marks in this question. Not being able to recall the definition in part (a) is inexcusable. In part (b), mentioning the initial and final temperatures would earn 1 mark.

Question 8

Ideal gases

(a) State Boyle's law. (1 mark)

(b) Explain what is meant by an ideal gas. (1 mark)

(c) An ideal gas has a volume of 0.67 m^3 at a pressure of 2.1×10^5 Pa. The temperature of the gas is 30°C. The gas is compressed so that its volume becomes 0.54 m^3 at a pressure of 4.5×10^5 Pa. Calculate the final temperature of the compressed gas in °C. (4 marks)

(d) The mean translational kinetic energy of gas atoms on the surface of a star is 1.5 eV. Calculate the thermodynamic temperature of the gas. (3 marks)

Total: 9 marks

Candidates' answers to Question 8

Candidate A

(a) Boyle's law: the volume of an ideal gas is inversely proportional to the pressure. The temperature must remain constant. ✓

Candidate B

(a) Boyle's law applies to a gas at a fixed temperature. It gives the relationship between the volume and pressure of the gas. That is: 'pressure × volume = constant'. ✓

This is a good start from both candidates. The answers may look different, but both are equally creditable.

Candidate A

(b) The molecules of an ideal gas do not exert inter-molecular forces on each other (unless they collide with something). An ideal gas obeys the ideal gas equation $pV = nRT$. ✓

Candidate B

(b) This is a gas at high temperature for which we can use the equation $pV = nRT$. ✓

Both candidates provide good answers.

Candidate A

(c) An ideal gas obeys the equation: $\frac{pV}{T}$ = constant. Therefore:

$$\frac{2.1 \times 10^5 \times 0.67}{(273 + 30)} = \frac{4.5 \times 10^5 \times 0.54}{T}$$ ✓✓

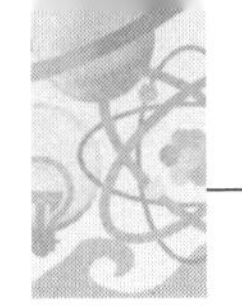

$T = 523.3$ K ✓

temperature = 523.3 – 273 = 250°C ✓

Candidate B

(c) initial temperature = 273 + 30 = 303 K final temperature = T = ?

$pV = nRT$

$$nR = \frac{2.1 \times 10^5 \times 0.67}{303} = 464.36 \checkmark$$

$$464.36 = \frac{4.5 \times 10^5 \times 0.54}{T} \checkmark$$

$T = 523$ kelvin ✓

Candidate A provides a perfect answer. The most common error in exams is working in °C rather than in kelvin. Candidate B does reasonably well, but it is a pity that the temperature of the gas was not changed back to °C — the candidate should have read the question carefully.

Candidate A

(d) mean kinetic energy, $E = \frac{3kT}{2}$

$$T = \frac{2E}{3k} = \frac{2 \times (1.5 \times 1.6 \times 10^{-19})}{3 \times 1.38 \times 10^{-23}} \checkmark\checkmark$$

temperature = 1.16×10^4 K ✓

Candidate B

(d) energy = $1.5 \times 1.6 \times 10^{-19} = 2.4 \times 10^{-19}$ joules ✓

I don't know how to find the temperature.

This is partly a synoptic question requiring knowledge of electronvolts. Candidate A continues to show excellent understanding and gives a perfect answer. Candidate B cannot answer the question but does well to get a mark for converting the electronvolts into joules.

Candidate A scores full marks. Candidate B scores 6 marks but is defeated by the last part of the question. Failure to select the equation for the mean kinetic energy of the atoms loses this candidate 2 valuable marks.